P9-BTO-916

Planet Earth

GRASSLANDS AND TUNDRA

Other Publications:

YOUR HOME
THE ENCHANTED WORLD
THE KODAK LIBRARY OF CREATIVE PHOTOGRAPHY
GREAT MEALS IN MINUTES
THE CIVIL WAR
COLLECTOR'S LIBRARY OF THE CIVIL WAR
THE EPIC OF FLIGHT
THE GOOD COOK
THE SEAFARERS
WORLD WAR II
HOME REPAIR AND IMPROVEMENT
THE OLD WEST

For information on and a full description of any of the Time-Life Books series listed above, please write:
Reader Information
Time-Life Books
541 North Fairbanks Court
Chicago, Illinois 60611

This volume is one of a series that examines the wonders of the planet earth, from its landforms, seas and atmosphere to its place in the cosmos.

Cover
A golden expanse of pristine grassland, one of the rare surviving natural tracts, covers the hills of western North Dakota. So fertile are the world's grasslands that most large areas have been converted to croplands: The grains raised there constitute about half the foodstuffs consumed by human beings.

Planet Earth

GRASSLANDS AND TUNDRA

By The Editors of Time-Life Books

Time-Life Books, Alexandria, Virginia

Time-Life Books Inc.
is a wholly owned subsidiary of

TIME INCORPORATED

FOUNDER: Henry R. Luce 1898-1967

Editor-in-Chief: Henry Anatole Grunwald
President: J. Richard Munro
Chairman of the Board: Ralph P. Davidson
Corporate Editor: Jason McManus
Group Vice President, Books: Reginald K. Brack Jr.
Vice President, Books: George Artandi

TIME-LIFE BOOKS INC.

EDITOR: George Constable
Executive Editor: George Daniels
Editorial General Manager: Neal Goff
Director of Design: Louis Klein
Editorial Board: Dale M. Brown, Roberta Conlan, Ellen Phillips, Gerry Schremp, Gerald Simons, Rosalind Stubenberg, Kit van Tulleken, Henry Woodhead
Director of Research: Phyllis K. Wise
Director of Photography: John Conrad Weiser

PRESIDENT: William J. Henry
Senior Vice President: Christopher T. Linen
Vice Presidents: Stephen L. Bair, Robert A. Ellis, John M. Fahey Jr., Juanita T. James, James L. Mercer, Joanne A. Pello, Paul R. Stewart, Christian Strasser

PLANET EARTH

SERIES DIRECTOR: Gerald Simons
Designer: Raymond Ripper

Editorial Staff for *Grasslands and Tundra*
Associate Editor: Sally Collins (pictures)
Text Editor: Thomas H. Flaherty Jr.
Staff Writer: Rita Thievon Mullin
Researchers: Patti H. Cass, Roxie France (principals), Scarlet Cheng, Jean B. Crawford
Editorial Assistant: Lori A. Johnson
Copy Coordinator: Kelly Banks
Picture Coordinator: Renée DeSandies

Special Contributors: Oliver Allen, Ronald H. Bailey, Karen Jensen, Brian McGinn, Gail Cameron Wescott (text), Martha Reichard George (research)

Editorial Operations
Design: Ellen Robling (assistant director)
Copy Room: Diane Ullius
Editorial Operations: Caroline A. Boubin (manager)
Production: Celia Beattie
Quality Control: James J. Cox (director)
Library: Louise D. Forstall

Correspondents: Elisabeth Kraemer-Singh (Bonn); Margot Hapgood, Dorothy Bacon (London); Miriam Hsia (New York); Maria Vincenza Aloisi, Josephine du Brusle (Paris); Ann Natanson (Rome). Valuable assistance was also provided by: Di Webster (Auckland); John Dunn (Melbourne); Felix Rosenthal (Moscow); Carolyn Chubet, Christina Lieberman (New York).

First printing. Printed in U.S.A.
Published simultaneously in Canada.
School and library distribution by Silver Burdett Company, Morristown, New Jersey.

Library of Congress Cataloguing in Publication Data
Main entry under title:
Grasslands and tundra.
(Planet earth)
Bibliography: p.
Includes index.
1. Grassland ecology. 2. Grasslands. 3. Tundra ecology. 4. Tundras. I. Time-Life Books. II. Series.
QH541.5.P7G727 1985 574.5'2643 84-28072
ISBN 0-8094-4520-4
ISBN 0-8094-4521-2 (lib. bdg.)

THE CONSULTANTS

James Estes was raised on the mesquite grasslands of Burkburnett, Texas. He received his B.S. from Midwestern State University and his Ph.D. in Plant Systematics from Oregon State University. Currently Estes is Professor of Botany and Curator of the Bebb Herbarium at the University of Oklahoma; he is also President of the American Society of Plant Taxonomists. He co-edited the volume *Grasses and Grasslands* and is now writing studies of the grasses of Oklahoma and the Southeastern United States.

Ralph E. Brooks is Assistant Director and Curator of the University of Kansas Herbarium and Botanist of the Kansas Biological Survey. He has conducted field work throughout the Great Plains from Texas to Canada and from the Rockies to the eastern forests. His studies of the flora of the Great Plains have appeared in more than 50 publications, including the *Manual of the Great Plains Flora,* a volume he also helped edit.

James Kenneth Detling is Associate Professor of Range Science at Colorado State University. Coauthor of more than 60 publications, technical reports and abstracts, he has won grants or awards for a number of them. His areas of special interest include grassland ecology, plant-water relations and plant-herbivore relations.

Dr. Stephen F. MacLean Jr. is Professor of Zoology and Head of the Department of Biology, Fisheries and Wildlife at the University of Alaska at Fairbanks. Professor MacLean received his Ph.D. from the University of California at Berkeley in 1969. He since has studied aspects of the tundra ecology in Arctic Alaska, Canada, Siberia and Finland and has published a number of papers that deal with tundra birds, lemmings, insects and other soil invertebrates. His current research focuses on the effects of temperature upon the ecological battle between tundra plants and the insects that feed upon them.

Samuel J. McNaughton is Professor of Botany at Syracuse University and a specialist on the plant ecology of Africa. Many of the 20-odd studies he has written or co-authored are primarily concerned with the Serengeti region of Tanzania and Kenya, whose grasslands and grazing ecology he has been studying since 1974 under a grant from the National Science Foundation. He has lectured widely and served as adviser on several ecological research projects.

Dr. Paul G. Risser is Chief of the Illinois Natural History Survey and Affiliate Professor of Plant Biology at the University of Illinois. He is the author or co-author of three books: *Man and the Biosphere, The True Prairie Ecosystem* and *Ecology and Natural History of the Plains and Prairies.* He has contributed to dozens of publications, and he has taken a leading role in 43 research projects involving ecological questions, natural-resource management and science education.

CONTENTS

A WORLD OF GRASSY PLAINS

A winter dust storm closes in on a caravan plodding through the parched grasses of the Eurasian steppe in northeastern Afghanistan.

Low grasses carpet the foothills of the Andes in central Colombia. This region is an equatorial savanna where the dry season lasts for six months and drought-resistant trees grow only in clusters near rivers and in thin ribbons along the beds of seasonal streams.

A weathered escarpment rises above acacia trees and tussocks of spinifex grass on the Australian savanna. This semi-arid region of Queensland once was an inland sea; now it receives so little precipitation that only the hardiest vegetation survives there.

An impending deluge darkens the sky over an acacia grove in East Africa's Serengeti National Park. The spectacular downpours of the Serengeti's rainy season provide enough moisture to support a scattering of trees and a rich cover of savanna grasses.

An early-evening rainbow follows a welcome September shower on the high plains of Colorado, a part of the North American grasslands that is subjected to extremes of weather, from savage blizzards in winter to summer temperatures over 100° F.

Dwarf birches and willows produce a crazy quilt of fall colors along a stream in the tundra of the Yukon. The changing shades signal an end to the tundra's brief growing season, which concludes in mid-September with the first snows of a winter seven months long.

Chapter 1

THE BURGEONING BOTANICAL FAMILY

In 1820, Major Stephen H. Long of the United States Army was sent west with a surveying team to study the geography of the region between the Mississippi River and the Rocky Mountains. He found an immense, treeless grassland — terrain so empty and so presumably poor that he pronounced it "uninhabitable by a people depending on agriculture for their subsistence." Further, he gave the region a bad name that persisted for decades: the Great American Desert.

Major Long was dead wrong, of course. The grassy plains of the American West, along with other grasslands of even more forbidding aspect, proved to be both habitable and incredibly bountiful. American grasslands now grow annual crops worth nearly $150 billion — plenty to feed nearly 250 million Americans, with enough left over to export vast amounts to less happily endowed nations. Indeed, crops grown on grasslands around the world satisfy three fourths of all human energy needs; and they produce still more food indirectly by providing animal forage, whose yield reaches the table in the form of meat, dairy products, corn-fed poultry and eggs. Fully 70 per cent of all harvested crops are classified as grasses, including wheat, corn, rice, barley, millet, sorghum and sugar cane. Still other species of grass, notably bamboo and giant reed grass, are used to make such diverse products as furniture, roof thatching, rope, hats and perfume essences.

Moreover, the grasslands have made historical contributions as vital as foodstuffs and raw materials. Thousands of species of flora and fauna developed special adaptations to their grassland environment. Among the animals were human-like primates that emerged from the forests of Africa about two million years ago. Features that proved advantageous to their life on the grasslands were an upright bipedal (two-legged) stance, which permitted ease of movement; an expanded field of vision; and hands with a thumb opposite the other four fingers, positioning that gave these primates the dexterity to hurl rocks and other weapons while hunting game. These changes, together with improved hand-eye coordination, ushered in the cultural and social evolution that came to be called civilization. After evolving together for many millennia, humankind and the grasslands are now virtually interdependent.

Yet in spite of their importance, the grasslands of the world are less well known than other major landforms. They have not aroused as much geological interest as mountains, possibly because they are not as dramatic or as immediately challenging. Nor have the grasslands elicited as much botanical interest as forests, perhaps because they are not as obviously diverse and

Acres of wheat, a domesticated grass, are tended by Egyptian workers in this detail from a tomb painting created around 1400 B.C. In the top panel, surveyors measure a field of grain with a rope tied at regular intervals. In the lower panels, the harvest is brought in and tabulated by scribes, one of whom sits on a pile of threshed grain and counts on his fingers. By the time these scenes were painted, cultivated grasses had been staples in the diet of ancient societies for at least five millennia.

picturesque. But grasslands are no less dynamic than mountains and forests. They expand, contract and even disappear with changes in climate. They can be created and destroyed — sometimes with appalling speed — by the radical changes that modern societies work upon the environment.

Today, grasses are the third-largest botanical family; only the orchid family and the aster family of flowers include more species. The Poaceae — the scientific name for the grasses — are divided into approximately 15 tribes. One of these is the wheat tribe, which includes wheat, barley and rye. Others are the fescue tribe, from which come a large number of lawn grasses; the rice tribe; and the oat tribe. The tribes in turn are subdivided into genera — about 600 in all — and then into species, of which 7,500 have been identified.

Counting only the large, unbroken grassland areas of the world, grasses cover almost 18 million square miles, or about 32 per cent of the earth's land area. The principal grasslands include the Great Plains of North America, stretching from Texas to mid-Canada; the Eurasian steppe, which extends from Hungary eastward through the Soviet Union to Mongolia; the lake- and lagoon-studded llanos of central and southern Brazil; the floor-flat pampas of Argentina and Uruguay; the vast and varied Australian grasslands; and Africa's tree-dotted savanna, which supports the last great natural preserves of wild animals. Akin to these grasslands are enormous tracts of tundra far to the north. Grasses also cover innumerable smaller areas and grow in more kinds of environments than any other botanical family.

Most of the major grasslands lie in the interiors of continents. They are characterized by continental climates, which feature wide daily swings in temperature and frequent extremes in weather — blizzards and tornadoes, droughts and dust storms. On the high steppes of Mongolia, for example, water sometimes freezes overnight as late as July, and during the day temperatures often rise to 85° or 90° F.

The location and extent of grasslands are affected by many factors, among them the character of the soil, the topography, and the altitude and latitude of the area. But climate is by far the most important influence. Generally, grasslands require from 10 to 40 inches of annual rainfall to survive. Lands that average less than 10 inches of rainfall usually become deserts, no matter how rich their soils, while more than 40 inches of rainfall fosters the growth of forests. Not surprisingly, grasslands are categorized and basically described by the climate zone in which each lies: Tropical, Temperate or Arctic.

The grasslands of these three climate zones display certain primary differences. The hot Tropical and sub-Tropical regions have a type of grassland known as savanna, whose plains are dotted with trees or shrubs. The annual rainfall in the savannas can measure as much as 47 inches a year, but the weather is ordinarily much drier, with three to seven rainless months every year. The Temperate Zone grasslands generally receive more precipitation — and more reliable precipitation. But in the Temperate regions the climate is harsher, with a much greater range in both annual and daily temperatures. In the cold domains of the Far North, great expanses of tundra support grasses and their near relatives, such as sedges; however, the principal plant forms in most areas are mosses and lichens. Like the other types of grassland, the tundra owes its special pattern of vegetation at least

A network of bamboo scaffolding, strong enough to withstand typhoon winds that collapse self-supporting steel scaffolds, is lashed to the façade of a building under construction in Hong Kong. The bamboo used here is one of about 1,000 species of giant grass. One species grows as much as four feet in a day, and another species reaches a height of 120 feet.

partly to limited amounts of moisture. Actually, some sections of tundra receive ample precipitation; however, much of the moisture remains unavailable to plant life because the ground is frozen for most of the year.

The wide distribution of grass plants has taken place in relatively recent times, for grasses are young as a botanical family. (Conifers, for example, are much older, having emerged 195 million years ago in the Mesozoic era, when most of the earth was covered by forests.) Pollen somewhat similar to that of grasses has been identified in fossil form in deposits dating back 65 million years, but no true grasslands seem to have existed in that early epoch.

The Global Range of Natural Grasslands

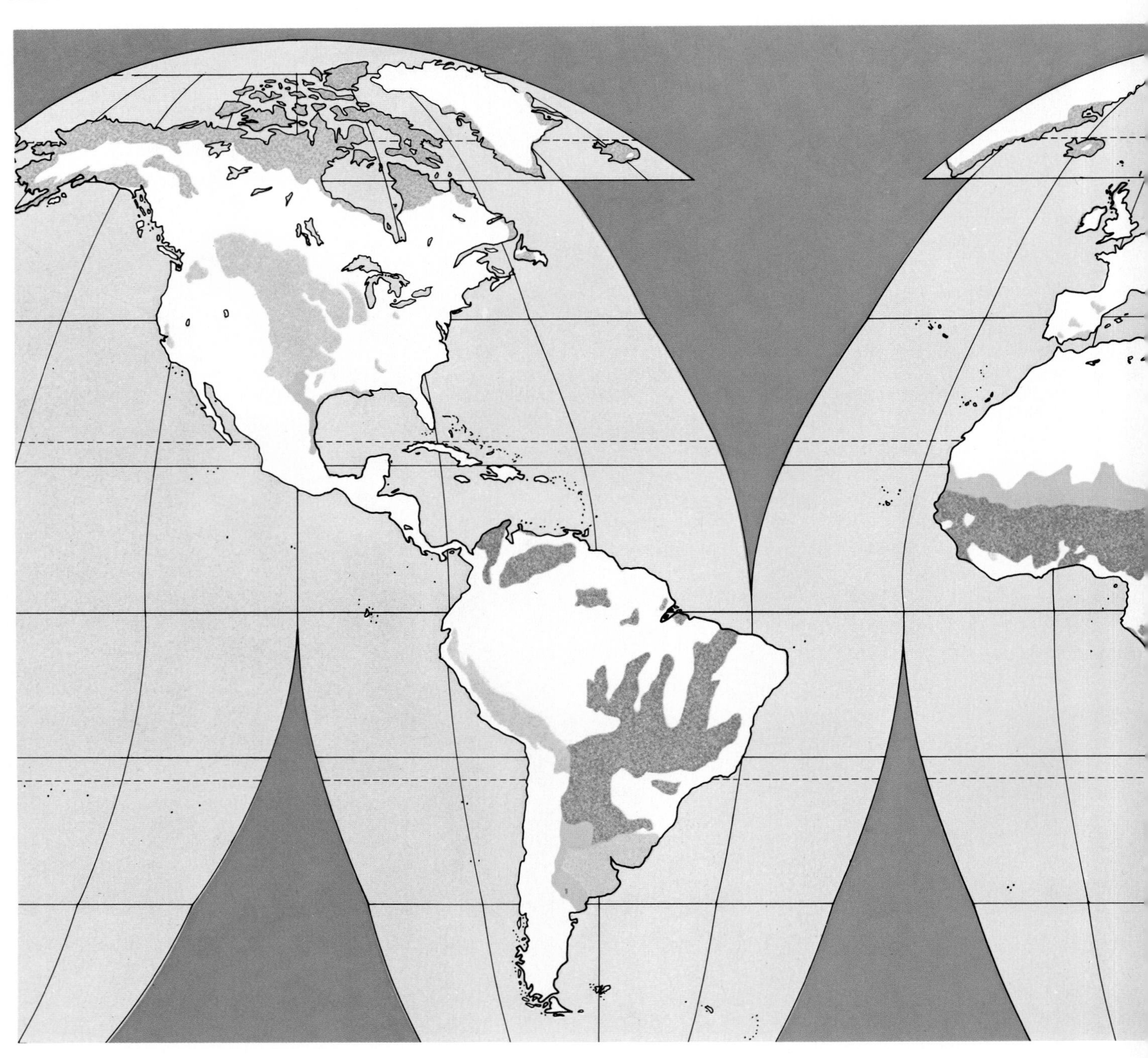

This map locates and characterizes the world's major types of grasslands, which together occupy 32 per cent of the land area covered by vegetation. Grasslands are transitional zones between deserts, which receive less than 10 inches of average annual precipitation, and forests, which get more than 40 inches of rainfall. Because the borders of grasslands fluctuate widely with changes in climate and in cultivation by humans, the map shows idealized natural grasslands — areas that scientists believe would be covered with grasses today if left untouched by human activities.

The five categories of grasslands are shaped by many factors. The patchy grasslands are interspersed with other areas because of broken topography and varying soils. Different patterns of rainfall account for two distinct types of savanna. Tree-dotted savannas have three dry months and seven or eight months of significant rainfall. By contrast, savannas that support shrubs have three very wet months and seven or eight dry months.

Among the Temperate Zone grasslands, the rainfall pattern is geographic as well as seasonal. In the North

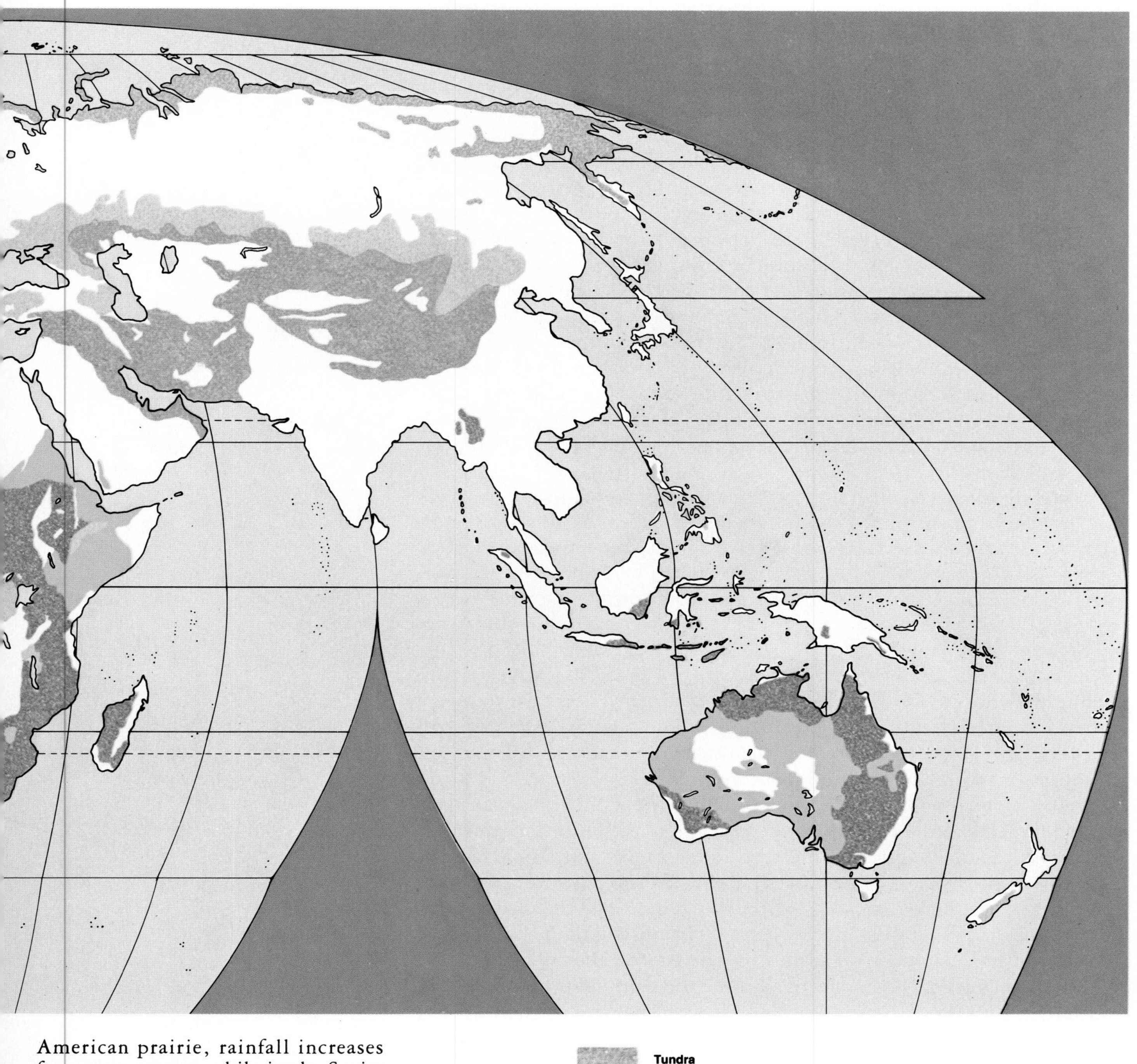

American prairie, rainfall increases from west to east, while in the Soviet Union's steppeland, precipitation increases from south to north. And in a conspicuous exception to the grasslands' limited rainfall, the cold tundra regions of the Far North receive abundant precipitation. But much of it is permanently unavailable, locked in an underground layer of ice.

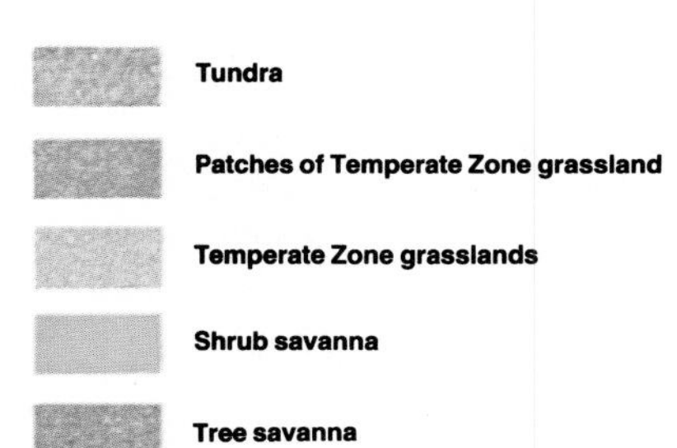

However, the global climate grew colder and drier during the Cenozoic era, roughly 30 million years ago. In this period, there was a general uplifting of the continents, and in some coastal areas, mountains rose to block the humid ocean breezes and to deny the inland regions as much moisture as they had previously received. As a result, great forests in the interior began to shrink and thin out. In addition, the existing soil was enriched by layers of topsoil washed down from the mountains. Grasses were admirably suited to this environment, and they gradually replaced the forests. By about 20 million years ago, grasses had become a major part of the earth's vegetation.

As the forests gave way to grasslands, the indigenous wildlife underwent compensating changes. Animals that had been browsers, feeding on the leaves and fruits of trees and shrubs, evolved and gave rise to grazers, which feed on grass. This transformation overtook the earliest member of the horse family, the Eohippus, or dawn horse. Eohippus was a browser the size of a dog, with toes and claws on its feet. But as the environment changed, the derived species grew larger; its central toe became a hoof, while the lateral toes shrank to become vestigial; its snout was elongated, with its lips becoming fuller and more mobile. Its teeth developed higher crowns, which were better suited to cropping grass leaves heavily impregnated with hard silica.

Paralleling the development of the horse family was the rise of the Bovidae, cud-chewing mammals specially adapted to life on the grasslands. Comprising such diverse creatures as cattle, sheep, antelope and gazelles, the Bovidae developed hoofs, high-crowned teeth and several other features that aided their survival as grass-eaters in open country. The females gave birth quickly, and within minutes, the offspring were ordinarily able to run and thus flee from danger.

The ruminant's chief adaptation was its specialized digestive system. Lacking enzymes to break down the fibrous walls of grasses, the ruminant evolved a complex four-chambered stomach whose cellulose-digesting bacteria permitted the animal to get at the calories, or energy, locked inside the cell walls of grasses. The ruminant bites off a large quantity of forage and gulps it down without benefit of much chewing. This mouthful goes into the rumen and the reticulum, the first two of the four stomach chambers: These chambers serve, in effect, as fermentation vats where bacteria break down the ingested plant matter. Later, the animal regurgitates the softened wad for further chewing. After this reworking, the food is swallowed again and passed through the first two chambers into the final two — the omasum and abomasum — as the digestive process is completed.

About two million years ago, long after the ruminants appeared, the woodland primates made their grassland debut. That they thrived in their new habitat has been ascribed by some scholars to what is called the edge effect; on the borderland between forest and open plain, there exists an environment rich in the flora and the fauna of both communities — and rich in opportunities. Early humans, standing at the edge of the woodlands and looking out over the flat grassland, could see distant herds of grazing animals. Stirred by the prospect of tender meat to add to their diet of fruits, nuts and roots, primitive peoples developed weapons with which to hunt effectively. They also learned how to use fire to flush prey from cover.

In the normal course of gathering food, early peoples found that the grain produced by grasses was an excellent source of energy and could be easily

A long bluff of loess rises from a cultivated grassland on the western border of Iowa. The loess, a fine soil 60 to 90 feet deep, was deposited here by the wind, which sifted it from the debris that a glacier carried down from the North during the last ice age, about 10,000 years ago. In time, water erosion carved narrow channels in the bluff. Trees grew in the moist channels, while the drier hillocks were covered with grasses and other prairie plants.

stored. This discovery led to a momentous practical application. Sometime between 8,000 and 10,000 years ago, hunters and food gatherers managed to cultivate certain crops by sowing collected grain in convenient fields, assuring themselves of a steady supply of food without traveling widely to obtain it. And they learned to domesticate wild cattle, goats and sheep, thus adding meat to the food supply they could obtain locally. Increased productivity permitted a surplus population to live in cities and practice commerce as well as crafts and arts.

Early in this formative period, the grasslands in the Northern Hemisphere were radically changed by the last in a series of ice ages. As great glaciers advanced from the northern polar regions, all living things retreated south before them — animals, humans and the grasses as well.

As the great sheets of ice melted back around 10,000 years ago, they deposited their prodigious loads of gravel, clay and sand. The most finely ground material was caught by the wind and dumped in mantles up to 300 feet thick. This loose granular material, which is known as loess, was particularly rich in nutrients and made an ideal rooting medium for grasses. The grasses gradually reclaimed the territory that they had lost to the invasion of glacial ice.

Even after the last glacier's retreat, the grasslands continued to move under climatic pressure. A few thousand years ago, when the North American climate was considerably warmer than it is today, the grasslands extended well to the east of their present margins. Around the start of the Christian era, a cooling trend began, bringing more humid air, and the grass retreated to the west. Since then, short-term climatic changes have

Fifty Million Years of Shared Evolution

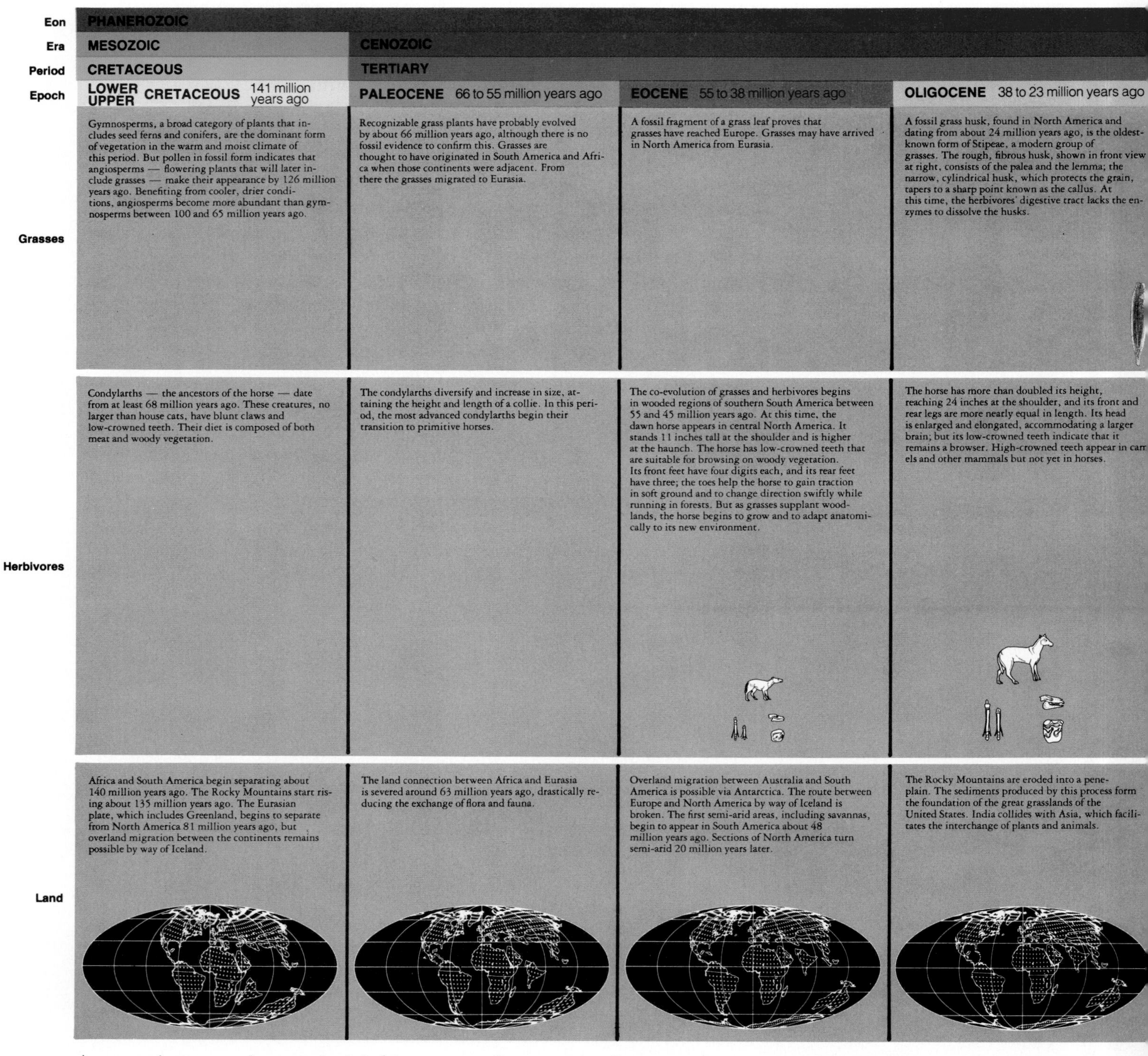

As an environment changes, its inhabitants evolve in response to the change — and to each other. This basic tenet of evolutionary theory has no stronger proof than the interrelated history of grasses and grazing animals. The chart above traces this relationship chiefly in terms of one group of grasses, the Stipeae *(top row)*, and one animal, the horse *(middle row)*, because their evolution is best documented by fossil evidence, most of which has been found in North and South America.

Long before grasses or grazing animals appeared in the record of life, the stage was being set for their emergence by profound geological changes *(bottom row)*. The continents, resting on immense plates in the earth's crust, were then clustered together but began drifting apart. In time, various land masses

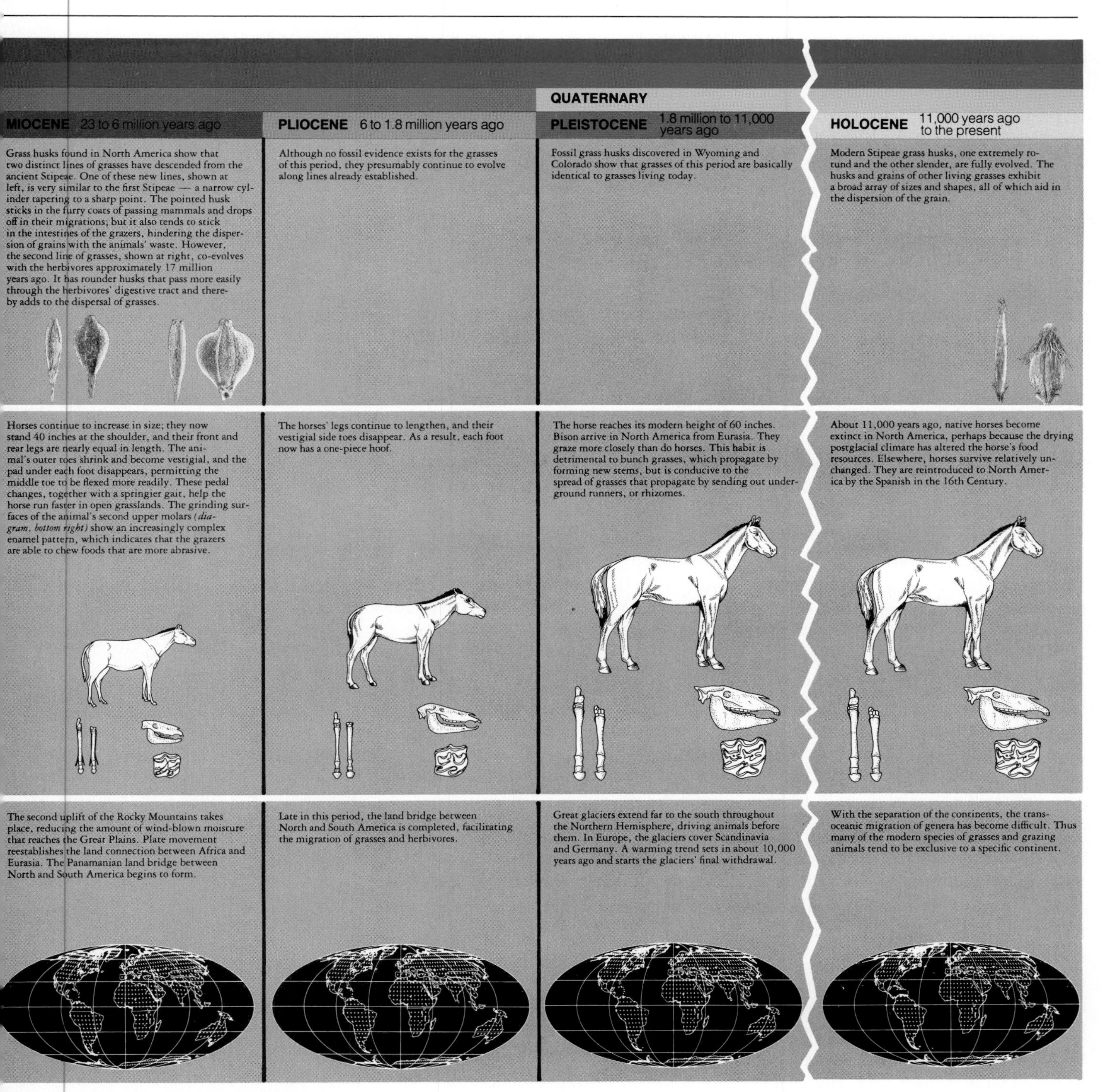

drifted back together and were connected by land bridges, such as the Isthmus of Panama. Naturally, the breaking of land links curtailed the migration of plants and animals, while the formation of new land connections not only increased the dispersion of species but led to many physical changes as these species adapted to their new environments.

By 50 million years ago, continental drift, together with the uplift of mountain chains, had altered the world's climate, creating drier conditions that favored the spread of grasses at the expense of forests. Stipeae and other grasses appeared and spread to form vast grasslands. Horses and other former forest dwellers began to adapt to their new environment and new diet. And the grasses, too, changed and diversified in response to the creatures that ate them.

occurred often, with commensurate shifts in the grasslands' perimeters.

Meanwhile, man-made changes in the location and extent of the grasslands increased more or less steadily. About 1,000 years ago, large tracts of Europe's forest cover were systematically removed as new feudal domains, striving for power, cut down trees to build towns and ships, to increase planted acreage for feeding their armies and to produce the charcoal needed for making steel weapons. Some scientists believe that much of the Russian steppe was once heavily wooded and that the forests were slashed to make room for cultivated crops.

In colonial America, settlers also cleared land for farming. New England, once fully forested under abundant rainfall, was 75 per cent fields by 1825. As Yankee farmers left for the West, much of New England reverted to its natural forested state. But the westering farmers quickly demolished most of what woods existed on their new acreage. By the beginning of the 20th Century, Ohio, once 90 per cent forested, was 85 per cent cleared fields.

In other areas, animal traffic created grasslands. Many open sections of England, for example, were once heavily forested. But many centuries ago, pigs, goats and cattle were for the first time allowed to forage in the woods for acorns, beech mast and other foods. As a result, the woodlands ceased to regenerate themselves and slowly dwindled, becoming extensive pasturelands carpeted with durable turf grasses.

The evolution of the grass family is still imperfectly understood. But scientists, working backward from the grasses' present form and distribution, have been able to make some likely deductions about how the plants evolved and dispersed after their first tentative appearance.

Though no one can yet say with any certainty where or when the first true grass species emerged, botanists speculate that it descended from a tropical plant not unlike one of the annual bamboo grasses of South America. Those low-growing, tufted plants with short oval leaves are quite different from the more common grasses that have tall, thin, narrow blades. But millions of years of evolution have brought about the transformation — and far more radical ones, at that.

Whatever their appearance, the early grasses spread and proliferated by what is known as adaptive radiation. In this process, the relatively unspecialized grasses gave rise to a host of species with highly specialized characteristics that evolved to meet a wide variety of local climatic and soil conditions. Wherever the grasses took hold, their genetic make-up changed in response to the requirements for survival. The species with the greatest ecological flexibility naturally became the most widespread.

Despite many diverse specializations, the grass family as a whole maintained certain common features *(diagrams, page 31)* besides its long, thin leaves. The leaves are attached to the stem in two alternating ranks, and the lower ends of the leaves — the sheaths — wrap around the stem like a split tube. The upper portion of the leaves — the blades — spread. The stem, or culm, is jointed with the base of the sheaths and is attached at each joint. The plant's small flowers, concealed by two scalelike bracts, are grouped into spikelets with two rows of clusters. The spikelets are bunched to form wispy brushes or feathery flags, usually found at the top of the long stem. And the grass plants have enormous root systems that may extend six

A typical prairie river, the Niobrara, winds its way

through Nebraska grassland. Wind-blown sediment from the river forms dunes whose sand and salt content contributes to the diversity of plant life.

Anatomy of a Grass Plant

The grass plant is a wonder of survival. The secret of its durability lies in its structure, which permits it to reproduce both vegetatively and by seed. The drawings at right illustrate the anatomy of a big-bluestem plant, one of the dominant tallgrass species on the North American prairie *(below)*.

The plant's stem, called the culm, is connected at intervals by joints known as nodes. Leaves alternate along the culm, with each leaf growing from a node. The lower portion of each leaf, the sheath, encircles the culm, and the leaf's upper portion, or blade, extends from the side of the plant. The leaf also includes a tiny appendage, called a ligule, where the sheath and the blade merge *(inset, far right)*.

Big bluestem, like many other grasses, may produce horizontal runners, or rhizomes, just below the soil surface. Other species have similar aboveground runners called stolons. Both rhizomes and stolons produce new plants from their nodes, thus forming a colony of plants without benefit of seeds.

The grass plant's seed-producing parts are more complex, but they are so small that they often go unnoticed. Typically, a magnifying glass is required to see the flowers. Each flower and its two protective scales, the lemma and the palea, are known as the fertile floret *(inset, near right)*. Big bluestem has a second floret that is sterile. Two additional scales, known as glumes, appear below the florets on the same tiny stem, or rachilla. Together the florets, the glumes and the rachilla are called a spikelet *(near right, top)*. A single plant may produce thousands of spikelets in one season. Collectively, the spikelets are known as the inflorescence.

A grass flower differs from other flowers in that two minute, colorless petals called lodicules extend from its base. When the flower is ready to open, the lodicules swell, forcing apart the lemma and palea and exposing the flower to the wind. Inside, each of the stamens includes a long filament bearing a pollen-producing anther, the output of which is astounding. A single anther on a rye plant, for example, produces 19,000 pollen grains.

The flower's pistil consists of an ovary and stemlike styles topped with feathery stigmas. The stigmas dangle from the flower in order to catch pollen carried on the wind. As the pollen penetrates the stigma, it grows down the style, fertilizing the ovule within the ovary.

After fertilization, the ovary matures into a grain. The collected spikelets *(far right)* begin to break apart, and both the spikelets and the grains they protect are released to the wind. The feathery awn protruding from the lemma of the big bluestem enables the spikelets to be wafted for considerable distances before they fall to the ground and germinate.

Maturing big-bluestem grass rises above the wild flowers on a South Dakota prairie.

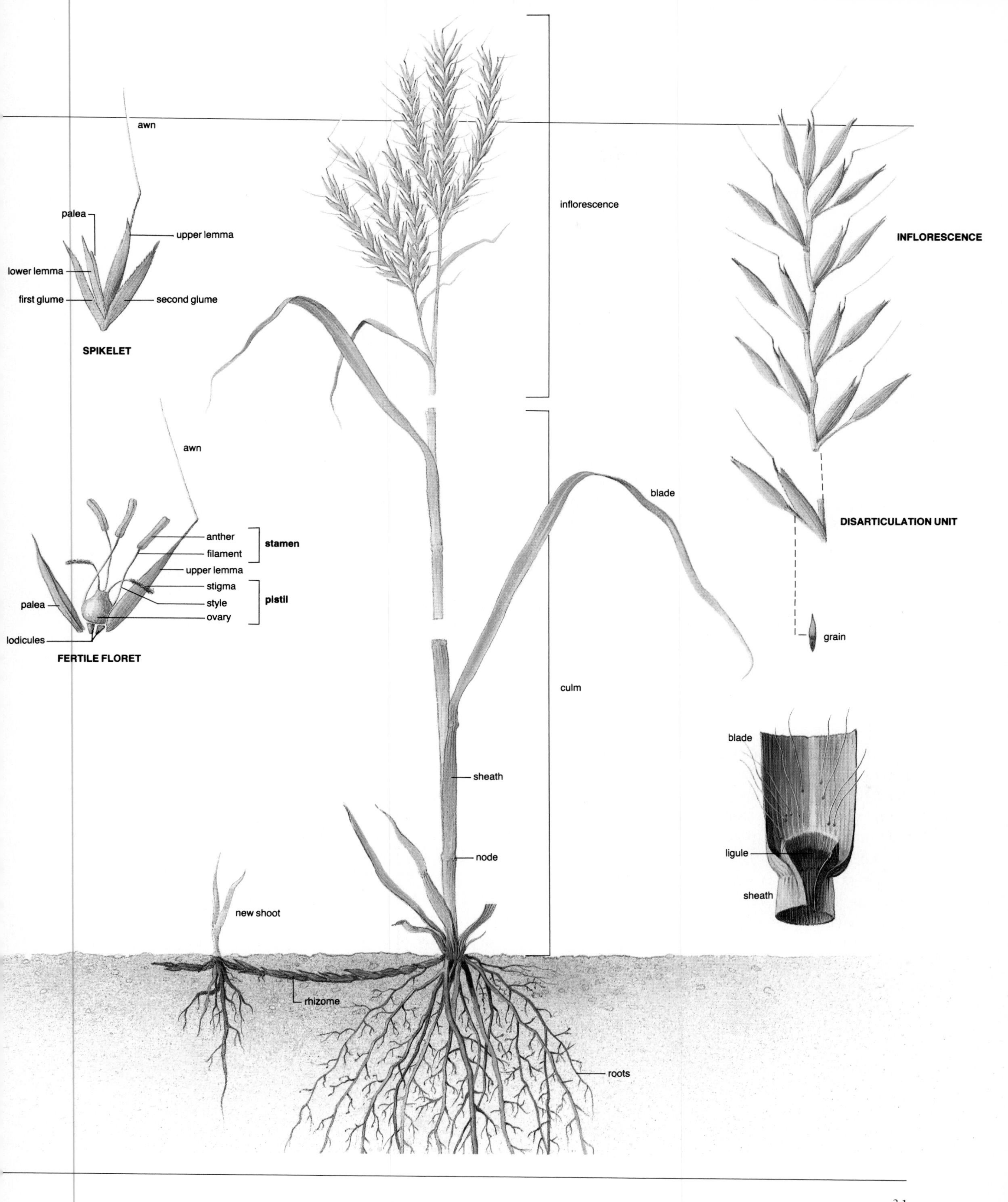
awn
palea
upper lemma
lower lemma
first glume
second glume
SPIKELET
awn
anther
filament
stamen
upper lemma
stigma
style
pistil
ovary
palea
lodicules
FERTILE FLORET
inflorescence
blade
culm
sheath
node
new shoot
rhizome
roots
INFLORESCENCE
DISARTICULATION UNIT
grain
blade
ligule
sheath

feet outward and even farther downward. It is not uncommon for a well-established plant to crowd two or three miles of roots and hairlike rootlets into its patch of ground. Importantly, this intricate meshwork of roots protects the soil from the forces of erosion.

One of the most impressive features of the grasses is their ability to survive conditions that kill most other plants: low rainfall, subzero cold, persistent desiccating winds, high temperatures, drought, burning, and the heavy grazing and trampling of herds. The key to the grasses' durability and resilience lies in their growth pattern. Cut a blade of grass, and it continues to grow — in the same way as human hair does after trimming. Grass stems also have growing points in a joint, or node, located just above the juncture of the leaf and the stem. Therefore the damaged stem of a plant can initiate new growth to replace the grazed tissue. Other growing points, called buds, are produced close to the soil surface.

The growing points can also produce new shoots known as tillers, which increase the number of stems per plant. Each new stem is capable of independent existence if the other stems are killed. Some grasses, such as big bluestem, reproduce in distinct clumps, with as many as 100 new stems growing from a single parent plant. This gives the plants a bunchy appearance, for which they are called bunch grasses.

Other species, known as sod-forming grasses, produce horizontal stems aboveground and belowground. From the tips and nodes of these appendages, which range in length from a few inches to a few feet, sprout roots and new aerial stems. The appendages that creep along the surface of the ground are called stolons, or runners. More often, the appendages extend underground and are called rhizomes; they thread their way like shallow roots through the upper four or five inches of soil. Both the stolons and the rhizomes send out stems and root systems of their own, and when any part of a sod-forming grass plant is destroyed, these offshoots will live on.

When the time comes for a grass plant to flower, it undergoes a number of changes. First the stem and other tillers elongate, ceasing to produce leaves at the apex. In the place of new leaves, the plant develops the inflorescence — the flower structure that will ultimately bear grains. Other stems may continue to put out vegetative growth but not flowers and seeds. The inflorescence is made up of a number of spikelets, each containing from one to more than 30 flowers. Until the flower is ready to distribute or accept pollen, it is tightly enclosed by two bracts, or chaff, which shield it from harm. When the flower has matured, organs at the base of the bracts swell with water and force the bracts apart, permitting the pollen to be blown away. Some of the pollen drifts to the flower of another plant, pollinating it.

Most grass plants have pollen-bearing anthers as well as pollen-receiving pistils; in these species, the pollen can fall from the anther to the stigma, and self-pollination does occasionally take place. But cross-pollination is more common. The pollen of one plant alights on another plant's pistil; the stigma, one part of the pistil, catches the pollen. Then the pollen germinates, and a pollen tube grows through the stigma into the ovary, where fertilization occurs. Not even one grain out of a thousand reaches the stigma. But the pollen is produced in copious amounts — as hay-fever sufferers know all too well. A single anther on a rye plant may contain more than 19,000 pollen grains. Since each rye inflorescence comprises at least 70

Drawings of the seed-bearing spikes of five kinds of wheat show three ancient species *(above)* and two modern wheats *(right)* derived from them by hybridization and chance mutation. The ancient species, probably the first wheats to be domesticated, are *(from left)* einkorn, emmer and *Triticum timopheevii;* their yields are low, and their spikes have a tendency to fall apart before harvesting, scattering the tough-shelled, hard-to-thresh seeds. The modern hybrids are durum (or macaroni) wheat and bread wheat. Spikes of these high-yielding species hold together better before harvesting, and their husks are more easily removed in the threshing process.

flowers, each with three anthers, one plant distributes about four million pollen grains. Of course, this enormous output of pollen increases the chances that grains will alight on receptive stigmas.

The marvels continue. The seed produced by the plant's ovary is really a fruit called a grain, which has two principal elements. One is the embryo, rich in protein and oil. The other is the endosperm, a storehouse of concentrated carbohydrates, protein, vitamins and minerals; these nutrients nourish the embryo until it can take root. Small wonder that the cereal crops produced by cultivated grasses are a beneficial fixture of the human diet.

The grains of some species have taken on novel configurations that improve their chances of getting planted. Grains of needlegrasses taper to a sharp point. This spearlike shape helps the fruit stick in the soil. And a remarkable apparatus enables it to penetrate farther. Each grain has at its apex a coiled moisture-sensitive filament. When dried by the air, the coil loosens, literally screwing the grain into the ground — or painfully into the hide of an animal.

A grain will lie dormant, thinly covered with soil, until the temperature and moisture levels are suitable for germination. Then the grain swells, cracking the scaly outer coat that protects it. As water enters, the embryo sends a root downward and a small shoot upward. The starch in the endosperm nourishes the growth — not directly, for starches are insoluble, but by converting the starch into soluble sugar. In a matter of time, the upward-growing stem breaks the surface. That is just the beginning, for a new perennial grass plant takes about five years to become fully established. Annual grass plants, such as fescue, germinate, grow, flower and set grain in just a few weeks of their short life cycle.

The perennials, which comprise fully 95 per cent of the grasses and cover most of the natural grassland, depend largely on their tillers, stolons and rhizomes for reproduction and bear relatively few grains. Since the annuals depend entirely on their grains for survival, they produce bumper seed crops. Their abundant production of seed has made them popular with growers since prehistoric times.

Because the cereals became staples of the human diet and were much in demand, farmers have sought for thousands of years to improve these crops' fertility and ease of harvest. The farmers worked particularly dramatic changes in the dominant grain, wheat, by first selecting preferable strains and later by hybridizing — combining strains to emphasize certain desired qualities. Flowering stems that were brittle in the wild plant, and therefore likely to shatter and let grains fall prematurely, have been made stronger and more supple. Grains have been induced to germinate more rapidly and to shed their chaff with less threshing. And the size of the grain has been enlarged.

Curiously, however, none of these early changes were as radical as those that nature brought about through chance hybridizations and mutations. The ancestor of all wheats is believed to be einkorn, or *Triticum monococcum,* a low-yielding species that originated in the Middle East. Einkorn is genetically a diploid — having seven pairs of chromosomes, or a total of 14. Einkorn readily hybridized with related species in the genus Aegilops (goat grasses), a group of weedy plants of the Middle East. Crossbreeding and chromosome doubling gave rise to another species called emmer wheat, which had 28 chromosomes. These tetraploids provided bigger yields, and

The beneficial effects of occasional fires are illustrated in this sequence of photographs, documenting a controlled burn of Missouri prairie. At the time of the burn *(top),* the grasses are dry and form a thick canopy; they burn quickly to the ground, as does a layer of dead organic material that intercepts rain water before it reaches the soil and also keeps the soil cool enough to discourage growth. Just 12 days later *(middle),* a scattering of vigorous new grass blades is already four to six inches high. Two months after the burn *(bottom),* the prairie has an abundance of thriving grasses.

some species had grains whose chaff peeled off readily under threshing. These wheats spread to Europe and Egypt. The most notable modern example of a tetraploid wheat is durum wheat, or *Triticum durum,* a hugely popular grain used to make pasta.

The tetraploids in turn hybridized with another goat-grass species, and another chromosome doubling occurred that produced a hexaploid, which had 42 chromosomes. The hexaploid species known as bread wheat, or *Triticum aestivum,* was extremely fertile, thanks to a mutation. This species has become the most widely grown strain in the world today.

At every stage of its existence, a grassland is considerably more than the sum of its grass plants. Each year, the great expanse of wind-rippled grasses modulates its colors from the opulent greens and blue-greens of new spring growth, to the reds, purples and greens of summer and finally to the burnished golds and reds of autumn. If the vista lacks detail to the naked eye, it is not nearly as empty and monotonous as it appeared to early travelers and landlookers; it is an eventful community where much takes place in three interrelated layers of life — on the surface, above it and below it as well.

Although wildlife is considerably limited on cultivated lands and fenced-in grazing ranges, almost any natural grassland or protected reserve supports an intricate web of relationships among creatures of many types and sizes — mammals, birds, insects and microorganisms. Each of these creatures in its own way subserves the health, integrity and dynamic balance of the grassland community.

On the surface is the herbaceous layer, consisting of the plants on which the big grazing mammals feed. On most grasslands with moderate moisture, dozens of different grasses grow; a square mile in Nebraska was once found to contain more than 200 species. Among the grasses, there are likely to be a number of grasslike plants, such as sedges, plus a sprinkling of herbs and other broad-leaved plants. The herbaceous layer has remarkable resilience. When lightning or spontaneous combustion starts a grass fire, the burning evidently stimulates new growth and increases its nutritional value.

The animals that live off the grasses differ greatly in kind and number from place to place. In their virgin state, the plains of North America had huge herds of just two species of large herbivores, the bison and the pronghorn antelope. The African savanna possessed — and still possesses — the greatest variety of herbivores, including the wildebeest, the gazelle, the zebra and the elephant. In sheer diversity of animal forms, the Eurasian steppe ranked second to the African savanna, boasting saiga antelope, wild cattle, horses, stags and boars. Other grasslands had a single dominant herbivore, such as Australia's kangaroo.

In return for their food, all these herbivores keep the grass vigorous by their grazing, and their droppings enrich the soil. The large herbivores also trample young trees on the margins of grasslands, which holds back the inroads of forests in well-watered areas. The grazers in turn become meals for the large carnivores, which leave enough scraps for the scavengers. The predators contribute to the ecosystem by thinning the herbivorous herds and thus preventing destructive overgrazing.

Occupying the lower part of the herbaceous layer are the small, burrow-

ing mammals, mostly rodents: mice, gophers, shrews, prairie dogs, hares and rabbits. Some of these species get about by hopping, a form of locomotion more efficient among tall grasses than running, and most species lead a nocturnal existence to elude such predators as foxes, weasels, owls and hawks. The burrowers consume weeds and roots in addition to grasses, and in return for their food, they cultivate the soil; digging mixes it, aerates it and makes it more porous for the passage of life-giving moisture.

Circulating above ground level in the grasslands community are a wide assortment of herbivorous birds and insects. In any large area, the number and variety of insects are almost beyond calculation: It has been estimated that as many as 1,000 insects may occupy a single square yard, and an average patch of grassland may contain more than 200 species. The total biomass, or bulk, of insects in any area is usually greater than that of the large mammals around them. The insect population can damage crops disastrously if it becomes too numerous, but normally it is kept in check by birds, small mammals and reptiles — and the farmers' insecticides.

The third ecological layer is beneath the surface, at the root level and below. It is likely to be a cool zone even on the hottest days; the thick grass cover shades the soil and reduces the sun's heat by as much as 20° F. Here, enormous numbers of flightless insects and invertebrates, such as grubs and worms, exist on grass roots and on smaller animals living in the soil. A single acre of grassland may contain up to 60 million springtails, a species of wingless insect that lives underground. The same acre may harbor three million earthworms, which not only aerate the soil but build more of it. They eat the organic matter that the microbes partially decompose, combine it in their guts with minerals ingested from the inorganic parent-rock material and then deposit this rich mixture on the surface of the ground. The great naturalist Charles Darwin, who was the first to call attention to the magnitude of this process, calculated that the earthworms in a single acre of grassland deposit about 15 tons of castings every year. Over the course of 50 years, the earthworms' excretions amount to the creation of 10 inches of new topsoil.

Numerically, the largest underground community consists of microorganisms — microflora, such as bacteria and fungi, and microfauna, such as protozoans. The bacteria within the top six inches of cultivated agricultural soil weigh anywhere from 290 to 640 pounds per acre. On natural grasslands, whose thicker and more extensive root systems support more microorganisms, the total weight of bacteria is likely to be far greater. The bacteria and fungi decompose organic materials in the soil. This process releases nitrogen, phosphorus and other nutrients, and makes them available to the roots of the grasses.

The ultimate contribution to the grasslands comes from the grasses themselves. Like all of the creatures that live off the grasslands, the grass plants die, decay and return to the soil everything they have taken from it — and more. Stems, leaves and especially the prolific roots decompose and are incorporated into the soil at the prodigious rate of close to half a ton in a year's time. Of course, this final contribution adds immeasurably to the productivity of the land.

Thus nearly all living things that benefit from the natural grasslands return important benefits to them. The balance in the grasslands community, as in all of nature, is wondrously perfect. Ω

A satellite photograph, at left, shows a patchwork of cultivated fields encroaching on virgin grasslands in a section of southwestern Australia. The farmlands are separated from the grasslands by a dark, irregular swatch of land, which is fenced off to prevent hordes of rabbits from invading and ravaging the croplands.

THE HARMONIOUS COMMUNITY

Plants and animals of the grasslands support one another through a complex web of relationships in which each contributes to the whole. Hundreds of animal species share the grasslands' bounty, utilizing almost all edible vegetation. Yet a natural system of checks and balances helps keep any species from dominating or destroying the environment. The animals affect their habitat by controlling plant growth, spreading seeds and recycling nutrients from plants to soil.

Each animal helps in its own way. Antelope and other hoofed beasts till the soil with every step, while burrowing rodents aerate and spread the subsoil. Elephants dig for underground water in dry periods, thus creating drinking sources for other animals. Browsing and grazing animals, or herbivores, consume part of grassland plants; the remaining tissues benefit from the available moisture and sunlight in the same way a suburban lawn thrives on regular mowing. East African herbivores have evolved a hierarchy of consumption. Zebras, the only grazers with teeth in upper as well as lower jaws, clear the coarse, taller layer of grass; wildebeests crop the grass a stage farther; little Thomson's gazelles nibble the lowest growth.

Plant life also plays a part beyond its passive role as a food source. The deep-rooted acacia tree, for example, draws nutrients from subsoil, and its falling leaves enrich surface soil, where grasses are rooted. The reproductive cycle of one variety of acacia makes use of the appetite of certain animals for its tasty fruit. Parasitic beetle larvae invade the seed-bearing fruit and, if undisturbed, eventually destroy the seeds. Once an impala or a gazelle eats the fruit, however, its digestive system kills the parasites and passes many seeds, undamaged, back to the earth to germinate.

On the East African savanna, a wildebeest tills the surface with its sharp hoofs, stirring up a meal of insects for the egret stalking patiently behind. Such tilling also creates niches where new grass seed can take root.

Two South American armadillos emerge from their subterranean home in the Brazilian savanna. The armadillos use powerful foreclaws to dig burrows, a practice that increases soil fertility by channeling oxygen underground and distributing fresh subsoils on the surface.

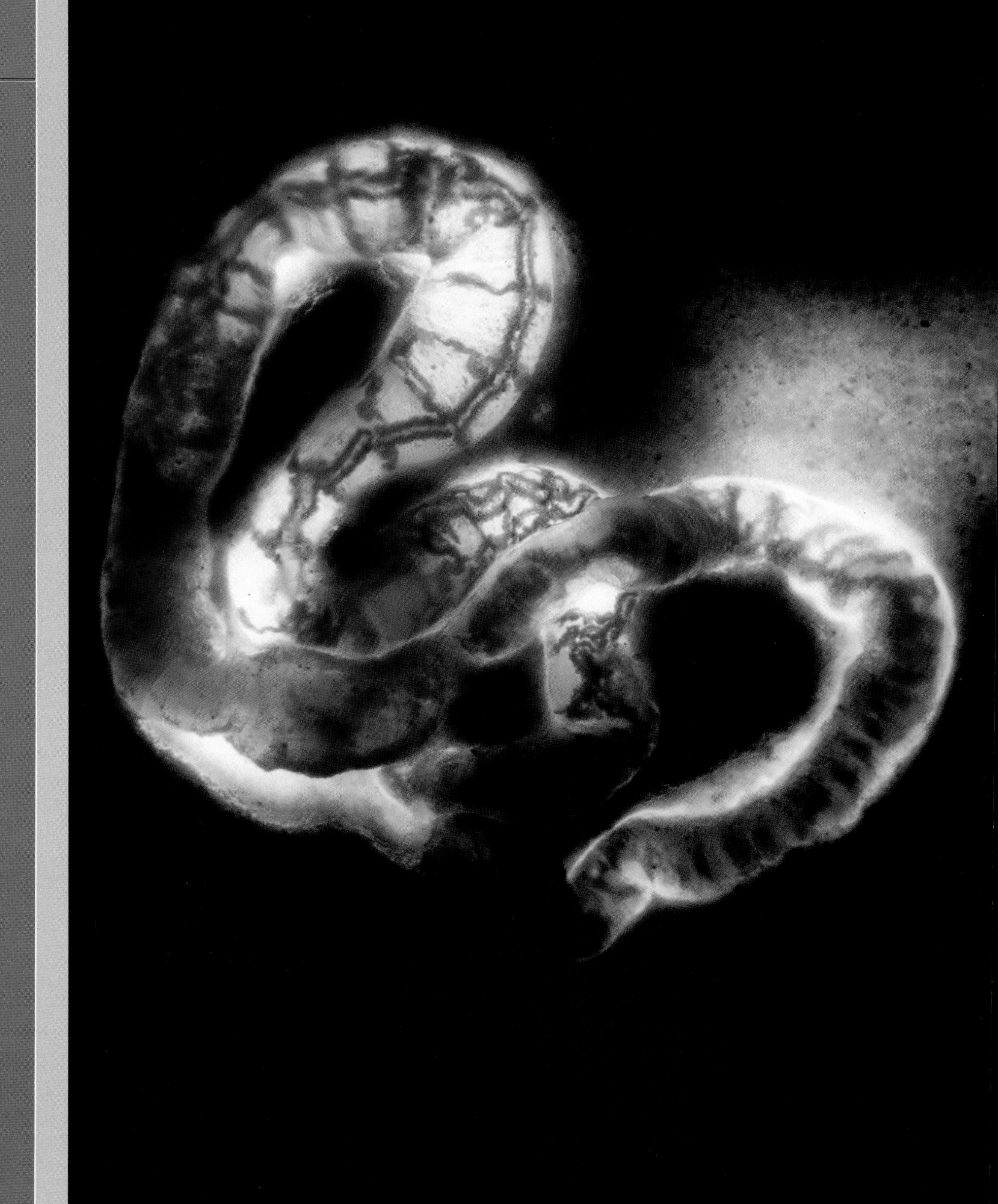

The American earthworm *(left)* is one of the most industrious tillers of grassland soils. The earthworms in a single acre pass an astonishing 15 tons of soil and vegetation through their digestive systems each year. In the process, organic matter breaks down into nitrogen, phosphorus and potash, which fertilize the grass.

The African dung beetle *(right)* recycles and redistributes the droppings of wildebeests and other savanna grazers, rolling the dung into one- or two-inch balls and storing them underground. Some of the dung's organic nutrients serve as food for the beetle's larvae, and the remainder is available as fertilizer for the soil.

Termite mounds on the Australian savanna serve as signposts to the sweetest, most nutritious stands of grass. The building practices of termites contribute to soil productivity: Their tunneling improves drainage, and the alkaline soils they bring to the surface from as deep as 12 feet neutralize acids in the upper soil.

Rolling on the ground to rid itself of insects, the American buffalo creates dusty, shallow depressions in which new seeds, formerly lodged in its hairy coat, may take root. The basins sometimes fill with rain, becoming ideal places for moisture-loving plants to grow, as well as drinking troughs for other animals.

Nibbling delicately on its favorite food, a giraffe browses among the tender leaves of an African acacia tree. The giraffe's constant pruning shapes the acacia's umbrella of foliage into a dense mass of leaves that creates midday shade for other animals.

Biting ants inhabit the hollow, bulb-shaped growths, or galls, at the base of the needle-sharp thorns on one acacia species. The ants and thorns fend off insects and small herbivores, but the giraffe's agile tongue and leathery mouth enable it to prune the tree with impunity.

A marsh hawk, which ranges the grasslands and swampy areas of the Northern Hemisphere, feasts on the innards of a freshly killed rabbit. By keeping rabbit populations in check, the hawk helps prevent such destructive practices as overgrazing and excessive burrowing.

Teaching her youngster by example, a kangaroo nibbles at the grass that covers almost half of Australia. Kangaroos and other herbivores often return to recently grazed grassland because the new shoots are more succulent and nutritious than older, more fibrous plants.

A lubber grasshopper, one of the 100 grasshopper species that inhabit American grasslands, is considered a voracious pest by ranchers. Yet the lubber eats only broad-leaved plants (its favorite is the tall sunflower) and, by controlling their growth, makes more sunlight and moisture available to the grasses below.

Chapter 2

PRAIRIE PRIMEVAL

Nothing they had seen in their European homelands prepared them for the immensity of it. First came the Spanish conquistadors under Francisco Vasquez de Coronado; in 1541 they rode north from Mexico in search of gold and penetrated perhaps as far as Kansas. Instead of gold, they found grass beyond imagining.

Like an endless sea it stretched to the horizon and beyond, interrupted only by thin ribbons of trees along riverbanks. "I reached some plains so vast that I did not find their limit anywhere that I went," Coronado reported, "although I traveled them for more than 300 leagues."

Coronado's men were astonished to find that their horses left no tracks in this ocean of green plants; the waist-high grass always sprang back, as straight as before. The land was so level and lacking in landmarks, wrote Coronado's chronicler, that "men became lost when they went off half a league. One horseman was lost, who never reappeared, and two horses, all saddled and bridled."

More than a century later, French explorers entered the North American grassland from the northeast. They emerged from forests into dense grass so tall that men on horseback had to stand in the stirrups just to see where they were going. In their effort to choose an appropriate name for this magnificent grassland, the explorers could do no better than adopt the French word for "farm field" or "meadow." The word was *prairie,* and it stuck.

Gradually, explorers and cartographers learned the full extent of the North American prairie. Roughly triangular in shape, it embraced 1.4 million square miles, or about 15 per cent of the continent. The base of the triangle extended north and south for 2,400 miles along the eastern side of the Rocky Mountains, from Canada's Northwest Territory to the present state of Texas. From this base, the arms of the triangle converged eastward to a point in Indiana more than 1,000 miles away.

The North American prairie was by no means the only grassland of its general type. Elsewhere in the world's Temperate Zones there existed other plains that received the right amount of precipitation — a range of about 10 to 40 inches annually — to support abundant grass, provided that soil and other variables were favorable. These grasslands, each named in the language of the dominant settlers, include the velds of southern Africa, the pampas of South America and, most extensive of all, the steppes of Europe and Asia, which stretch for more than 3,600 miles between Hungary and Mongolia. Smaller areas of temperate grasslands can be found in many countries, among them New Zealand, Turkey, Iran and West Germany.

Big bluestem, the dominant tallgrass of the eastern Great Plains, towers to heights of seven feet or more and reaches six feet deep into the soil for its nutrients. This archetypal prairie grass provides shelter and forage for wildlife ranging from deer and coyote to the 10-spotted dragonfly *(foreground).*

A rider crosses the treeless Mongolian steppe, which has been turned lush by spring rains. Nomadic herders follow the rains with flocks of sheep that monopolize the grass and water, threatening the steppe's wild grazers with extinction.

Nearly all of these grassland areas have been converted from their natural state into grainfields or are being used as stock-raising pasturage, and today they produce about 70 per cent of the world's foodstuffs.

Not much is known of the virgin grasslands of Asia, Europe and the Middle East. In the populous areas of the ancient world, these regions were put to agriculture long before they were studied systematically. But different circumstances made the North American prairie a historical treasure. The Indian populations of the American plains were always very small and did little to alter the environment. Thus most of the prairie remained in its natural state well into the 19th Century. By the time settlers pushed west in large numbers after the Civil War, naturalists were following closely, some of them actually examining acres of grassland on their hands and knees. As a result of their work and the continuing intensive study of the few unspoiled tracts remaining, the North American prairie is known in greater detail than any of the world's other grasslands. In turn, knowledge of the prairie has shed light on the dim early history of grasslands elsewhere.

The early naturalists saw a carpet intricately woven of many strands. The North American prairie — and indeed all Temperate Zone grasslands — can be divided into three main types according to the height of the grass. These classes are shortgrasses, measuring less than one and a half feet high; midgrasses, two to four feet; and tallgrasses, five feet or more. Though grasses of all three classes appear in nearly every area of the prairie, one class tends to control a given geographical region. Shortgrasses dominate in the western section, midgrasses in the central portion and tallgrasses in the east.

This geographical gradient reflects the local availability of moisture. The wet winds from the Pacific rise, cool and dump most of their moisture when they reach the Rocky Mountains. Thus the high plains just to the east fall under a so-called rain shadow; in a band up to 200 miles wide, annual precipitation averages as little as 10 inches, and strong winds quickly evaporate much of that. As a result, shortgrasses, such as blue grama and buffalo grass, dominate here.

Farther east, the prairie starts to emerge from the Rockies' rain shadow. Winds bring in moisture from the Gulf of Mexico and the northwestern Arctic region and, at about the 98th meridian in Kansas and Nebraska, annual precipitation rises above 20 inches. Here begins the mixed-grass prairie. As the name implies, the region is a transition zone possessing grasses of all three classes, but it is dominated by the midgrasses, particularly the species known as little bluestem and side oats grama.

Finally, about 400 miles farther east, begins the tallgrass region. The king of the tallgrass is big bluestem, a majestic species that can grow to 12 feet in height when it receives about 40 inches of annual precipitation.

Blessed by the prairie's most abundant rainfall, the tallgrass region contains the greatest diversity of native vegetation. A single acre of uncultivated land may harbor as many as 300 different species of plants. Big bluestem and a few associated grasses make up most of the ground cover. The rest are forbs — broad-leaved herbs, such as goldenrod and aster. Their flowers, blooming from April through October, bring to the tallgrass region a kaleidoscope of colors.

The many species of grass and forb can crowd into a relatively small space because of the prairie's system of vertical layering. Each species occupies its

The Long and Short of Prairie Grass

TALLGRASS

Adaptability and tenacity, traits that were essential to human survival on the North American prairie, are equally important in the three major groups of grasses — tallgrass, midgrass and shortgrass — that evolved there. Examples of the three groups are shown on these pages in scale. They vary strikingly in height, from eight feet to a mere four inches — a range that reflects each grass's opportunistic response to the amount of available rain.

Each group has its own quirks and characteristics. Tallgrasses, such as big bluestem, block sunshine to other grasses, while their own seeds thrive in near-darkness. Shortgrasses hug the ground in semi-arid areas and hide most of their growth under the surface. Midgrasses shift across climatic boundaries and thrive in niches too cold for shortgrass or too dry for tallgrass.

Indian grass grows as tall as eight feet, but it cannot spread horizontally — a liability that results in small, isolated stands of the golden grass.

Nutritious big bluestem is the grass most favored by cattle, especially in early bloom, before its stems become hard and fibrous. Big bluestem is dominant in the lowlands of the eastern prairies and in the Missouri, Platte and Red River valleys farther west.

The comblike spikes of cordgrass (shown here hosting a pai leaf beetles) reflect the coarse, serrated texture of this fast-growing tallgrass, which thrive on land too wet or too poorly ae ated for big bluestem.

MIDGRASS

SHORTGRASS

Little bluestem, a midsized grass, uses moisture more efficiently than big bluestem, its tallgrass relative. Thus little bluestem dominates the dry western prairies of North America and competes with big bluestem on dry slopes and uplands farther east.

Needlegrass adapts its growth pattern to the poor, gravelly soil and cool climate of the northern prairie. It sprouts from seeds that burrow deeply into the topsoil, and it matures with a rapid burst of spring growth.

Foxtail barley, a relative of the domesticated barley grain, grows in weedlike profusion on wet prairies. When foxtail barley is young, its bushy flowers provide grazers with tasty forage, but at maturity the flowers develop sharp, injurious spikes.

Buffalo grass survives by lying low, curling its leaves near the soil to preserve some of its growth from the bison's close cropping. This grass sends out aggressive, aboveground shoots that invade bluestem habitats heavily grazed or weakened by drought.

LITTLE BLUESTEM

NEEDLEGRASS

FOXTAIL BARLEY

BUFFALO GRASS

own niche aboveground and belowground. A forb such as button snakeroot eryngo (a member of the carrot family) may grow near the ground but thrust its flowering parts above the taller grass to ensure pollination and seed dispersal. Others may mature early in the season, reaching fruition before they are shaded by taller species. Similarly, forbs and grasses tend to send roots to different depths, which allows each species to obtain moisture from its own level. This ability to coexist results from evolutionary selection of genetic traits in response not only to climate and soil but also to the characteristics of neighboring species.

Even more remarkable is the evidence of differences among plants of the same species. One species of grama grass, for example, thrives in two sharply contrasting prairie habitats: in the north, where the growing season is limited to only 100 days, and in the southern fringes, where the season lasts more than three times as long. There are subtle genetic differences between these two plants of the same species. When grown in the laboratory, the northern variety matures much faster than its southern counterpart. It also responds better to the longer days characteristic of its home in the north.

In fact, genetic differences are also found between individual plants of the same species growing close together. These variations in genes are vestiges of the individual plant's evolutionary heritage, reflecting changes that have occurred in response to different origins or slightly different microclimates.

In growing densely aboveground and belowground, prairie grasses form a thick, tough mat of sod. Just how tough was attested to by the early pioneer farmers, who always spoke of the first plowing in their new grassland acreage as "breaking the prairie." In fact, the sod broke many wooden and even iron plows. The problem was finally solved in the 1840s, when a blacksmith named John Deere put a steel plow on the market. But the Deere plow, too, could fail to do its job; at least two oxen were needed to drag it through the most tangled sod mats.

After the plow cut through the labyrinthine growth, the sod still needed two or three years to rot before it became readily tillable. Pioneer farmers could not wait that long to grow their own food. They took an ax to the broken prairie, dented the bared soil, and here and there they dropped in a handful of corn seed.

Happily, the farmers discovered that the prairie sod made a serviceable substitute for lumber on the treeless plain. They sliced the sod into long, thick strips with the plow. Then they cut these strips into pieces about a foot wide and two feet long and stacked these sod bricks to form the walls of their houses. Sod houses of various types protected many a pioneer family from the fierce winds and cold winters of the open prairie.

The sod houses were usually built of big bluestem or its neighbor on the tallgrass prairie, slough grass. A coarse, tough-stemmed species that may reach nine feet in height, slough grass made firm sod but was difficult to handle because its leaves had sharp, saw-tooth edges — "rip-gut," oldtimers called it. Big bluestem was easier on the hands and also more abundant.

Prairie grass plants need their extensive root systems to anchor them, to absorb water and to obtain nutrients. Consistent with those needs, the little bluestem seedlings used in laboratory experiments first concentrated their energy into putting down roots. The seedlings, nurtured in optimum conditions without competition for nutrients and water, grew a primary root more than two inches long before the first aboveground shoot appeared. After just two weeks, its rapidly developing root system extended six inches deep and nearly half as wide. The aerial shoot, by contrast, was scarcely an inch high.

The subterranean parts of grass plants continue to outgrow the aboveground portions. By some estimates, roots and other underground appendages such as rhizomes make up more than 80 per cent of the vegetative biomass in a prairie. In fact, the living materials contained in just the upper four inches of soil in a typical acre of tallgrass weigh at least two and a half tons and perhaps as much as four tons. The roots of some grasses are so abundantly branched that the total growth in a square yard of soil four inches deep might, if placed end to end, stretch 20 miles.

Much of what is known about these extraordinary underground networks came from the research of John E. Weaver, an eminent plant ecologist, who spent most of his career at the University of Nebraska. Weaver labored for more than four decades to unravel the prairie's secrets. During his frequent field trips, he often had his students dig a big trench — 12 feet long, 3 feet wide and 7 feet deep — next to a plant under scrutiny. Then Weaver, still attired for the classroom in suit and tie, dropped into the trench and with an ice pick began the painstaking work of extricating the plant's elaborate root system from the wall of the trench.

Weaver's research indicated that the roots of tallgrasses, such as big bluestem, grow at remarkable speed — up to half an inch a day — and routinely penetrate to a depth of six feet. The roots of one species, switch grass, sometimes go down more than a dozen feet.

Root growth is typically adapted to reach the maximum depth of moisture penetration. The tallgrasses, growing in the prairie region of heaviest rainfall, have deeper roots than the shortgrasses of the semi-arid plains. Interestingly, however, the roots of such shortgrasses as blue grama and buffalo grass run deeper relative to plant height than do the roots of the tallgrasses. Seldom more than 10 inches high, these plants send down well-developed root systems three feet deep that provide reserves of moisture in times of drought.

In addition to their vital nurturing role, the root systems of prairie grasses provide for year-to-year survival of their plants in the face of extremes of heat and cold. Unlike trees and shrubs, grasses and forbs die back to the ground during winter. Shielded by their blanket of soil and sod, the roots and new buds for next year's growth lie dormant and safe through winter temperatures as low as $-40°$ F. at ground level. Similarly, the underground systems can withstand the raging heat of fire. Investigators have shown that during a prairie fire the temperature three feet above the ground may reach 400° F., while just an inch or two belowground it rises only a couple of degrees.

Like the prairie grasses, the soils that sustain them in the various temperate grasslands are marvels of nature's engineering. Pedologists — soil scientists — classify soils partly according to their characteristic colors, which usually reflect their content of organic matter. On the North American prairie, these colors tend to follow the gradient of shortgrass, mixed-grass and tallgrass regions eastward from the Rocky Mountains. The colors range from brown through chestnut to nearly black, generally growing darker and richer as the dominant grasses increase in height.

For nomadic Indians and white settlers alike, the American prairie was a natural pharmacy. They used the herbs and common weeds they found in the grassland's open expanses and shady oases to relieve the aches, illnesses and injuries incurred in their rugged outdoor life.

A majority of these folk remedies have never undergone rigorous scientific scrutiny, but modern pharmacologists have determined that a number of prairie plants do contain useful medicinal ingredients. Several of the herbs shown at right, for example, have an antiseptic effect on sores and cuts. Others have chemical ingredients that made those herbs capable of soothing gastric upsets and easing the pain of childbirth for Indian and pioneer women.

Some of these herbal medicines, however, were more dangerous than the diseases they were meant to remedy. Gentian, for instance, could stimulate a sluggish appetite — or bring on nausea and vomiting, the reaction depending on the dosage administered. Poison ivy, an Indian cure for ringworm, could cause a severe allergic reaction in patients sensitive to its potent resins.

In the case of certain herbs, the ritual and mystique involved in herbal medicine were more powerful than any ingredient in the plant. Wild columbine, considered an aphrodisiac by the Omaha Indians, worked its magic exclusively through the power of suggestion: Young warriors, after crushing columbine seeds in their palms, vied to hold hands with comely maidens.

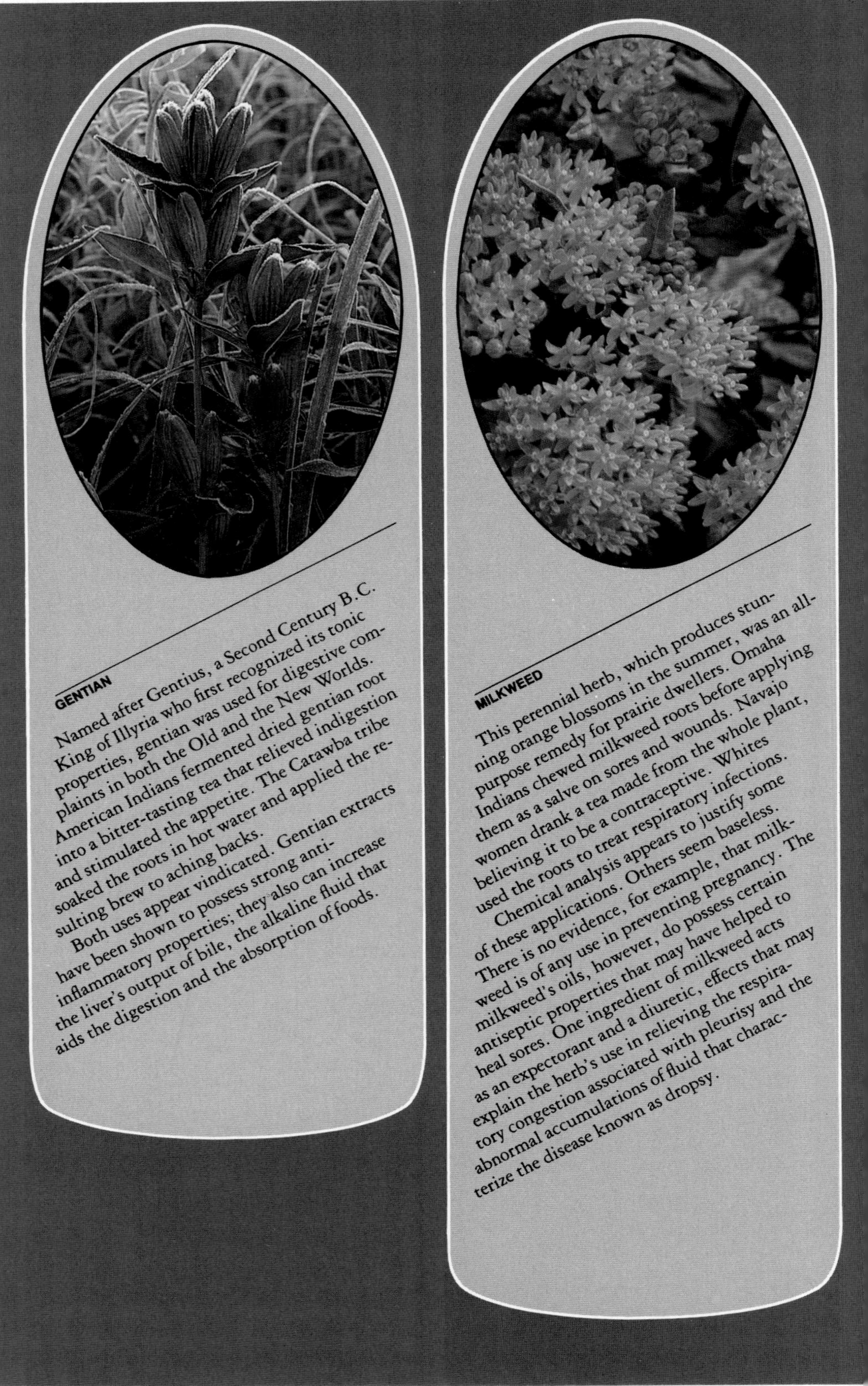

GENTIAN

Named after Gentius, a Second Century B.C. King of Illyria who first recognized its tonic properties, gentian was used for digestive complaints in both the Old and the New Worlds. American Indians fermented dried gentian root into a bitter-tasting tea that relieved indigestion and stimulated the appetite. The Catawba tribe soaked the roots in hot water and applied the resulting brew to aching backs.

Both uses appear vindicated. Gentian extracts have been shown to possess strong anti-inflammatory properties; they also can increase the liver's output of bile, the alkaline fluid that aids the digestion and the absorption of foods.

MILKWEED

This perennial herb, which produces stunning orange blossoms in the summer, was an all-purpose remedy for prairie dwellers. Omaha Indians chewed milkweed roots before applying them as a salve on sores and wounds. Navajo women drank a tea made from the whole plant, believing it to be a contraceptive. Whites used the roots to treat respiratory infections.

Chemical analysis appears to justify some of these applications. Others seem baseless. There is no evidence, for example, that milkweed is of any use in preventing pregnancy. The milkweed's oils, however, do possess certain antiseptic properties that may have helped to heal sores. One ingredient of milkweed acts as an expectorant and a diuretic, effects that may explain the herb's use in relieving the respiratory congestion associated with pleurisy and the abnormal accumulations of fluid that characterize the disease known as dropsy.

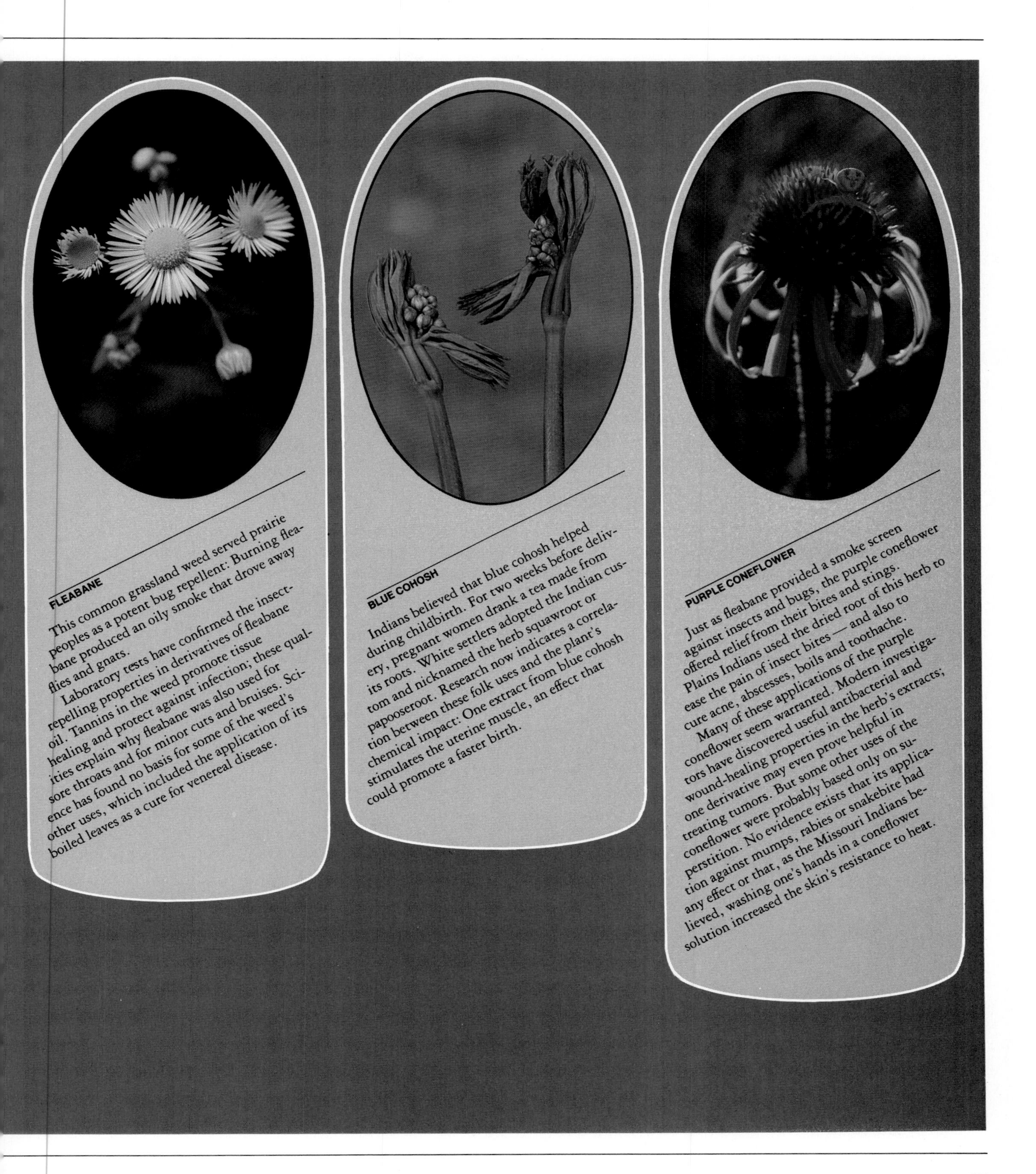

FLEABANE

This common grassland weed served prairie peoples as a potent bug repellent: Burning fleabane produced an oily smoke that drove away flies and gnats.

Laboratory tests have confirmed the insect-repelling properties in derivatives of fleabane oil. Tannins in the weed promote tissue healing and protect against infection; these qualities explain why fleabane was also used for sore throats and for minor cuts and bruises. Science has found no basis for some of the weed's other uses, which included the application of its boiled leaves as a cure for venereal disease.

BLUE COHOSH

Indians believed that blue cohosh helped during childbirth. For two weeks before delivery, pregnant women drank a tea made from its roots. White settlers adopted the Indian custom and nicknamed the herb squawroot or papooseroot. Research now indicates a correlation between these folk uses and the plant's chemical impact: One extract from blue cohosh stimulates the uterine muscle, an effect that could promote a faster birth.

PURPLE CONEFLOWER

Just as fleabane provided a smoke screen against insects and bugs, the purple coneflower offered relief from their bites and stings. Plains Indians used the dried root of this herb to ease the pain of insect bites — and also to cure acne, abscesses, boils and toothache.

Many of these applications of the purple coneflower seem warranted. Modern investigators have discovered useful antibacterial and wound-healing properties in the herb's extracts; one derivative may even prove helpful in treating tumors. But some other uses of the coneflower were probably based only on superstition. No evidence exists that its application against mumps, rabies or snakebite had any effect or that, as the Missouri Indians believed, washing one's hands in a coneflower solution increased the skin's resistance to heat.

The darkest, most fertile soils underlie the tallgrass region, which was enriched by thick deposits of topsoil brought south by the great glaciers of the last ice age. Most prominent are the types of soil that early Russian pedologists called chernozems — meaning "black earth." These soils, now known scientifically as mollisols but still popularly called chernozems, were first identified in the steppes, which now comprise some of Europe's most fertile farmlands. In North America, nearly half of the world's corn and much of its wheat and soybeans are produced in chernozem soils and in other rich soils in the eastern portion of the tallgrass region.

A vital ingredient in all prairie soils is humus — the partially decomposed organic matter that amateur gardeners eagerly concoct in their backyard compost piles. Humus represents up to 10 per cent of the content of chernozem and prairie soils and is responsible for their deep, dark coloration. (When humus is removed chemically from chernozem, the remaining soil is a pale gray.) Humus serves many purposes. As it decays, it slowly releases its mineral content in amounts that the grasses can utilize most efficiently. It keeps the grassland sods light and well aerated; humus-rich prairie soils consist of about 50 per cent air, which prevents compaction under a soaking rain and makes the surface springy under the heavy tread of humans or other large mammals. Humus also helps conserve moisture in the soil — a function of life-or-death importance during a drought.

Periodic droughts are a normal occurrence in temperate grasslands, and consequently many species of grass have evolved physical equipment to survive the dry spells. Big bluestem, despite its considerable requirement for water, endures by relying on its deep roots to draw moisture from the subsoil below. Relatively shallow-rooted shortgrasses, such as buffalo grass and blue grama, evolved in a drier climate and have developed resistant traits. For example, they curl their leaves to reduce evaporation, and they soak up moisture from the soil through a profusion of small hairlike fibers that jut from their roots.

Western wheatgrass and other species of midgrass survived by developing a drought-anticipating growth pattern. The plant gets a head start on the dry summer season by growing rapidly in the early spring, when moisture is normally ample. By late spring the plant matures, sends out its long reproductive rhizomes and produces a large number of seeds. Then the plant goes semidormant. And so it remains throughout the searing months of summer, waiting for the cool of autumn to resume growth.

The rapid early growth of these midgrasses increases their chances of survival in another way as well. The species soaks up the moisture in the soil, to the detriment of late-growing neighboring species. When the water-deprived neighbors succumb to drought, they leave more water, more nutrients and more bare prairie for the spreading of wheatgrass.

Such competition for survival is most evident in the transitional mixed-grass region. For example, during the worst dry spell on record — the seven-year drought that began in the American West in the early 1930s — the boundaries of the region shifted radically to the east in Nebraska, Kansas and Oklahoma. Quite naturally, the shortgrasses, being accustomed to little rainfall, gained dominance toward the east, where less drought-resistant grasses ordinarily prevail.

The shortgrass invasion was eventually reversed. The process began in 1941, when rainfall returned to normal. But it took more than a decade of

A bison bull raises a cloud of dust from the Montana prairie as it charges a rival during the annual mating season. Such churning and trampling by the one-ton bison ultimately improves the prairie's health; it clears away old growth, allowing new grass and herbs to take root.

normal precipitation before the midgrasses and tallgrasses again asserted their dominance in their respective regions.

With the development of prairie grasses came the evolution of a diverse bestiary of plant-eaters, ranging from the majestic bison to the common grasshopper. The mammals possess the usual adaptations to grasslands in general. The herbivores and the ruminants have durable, high-crowned teeth — and the ruminants also have special digestive systems — to break down the hard cell walls of grass and get at the proteins and carbohydrates contained inside.

The prince of the prairie mammals was the bison, or buffalo. When Coronado arrived in 1541, so many bison darkened the plains (by some modern estimates up to 60 million) that he found it "impossible to number them." Coronado's men and later the American pioneers were awed by the bison's size and might, which protected the great beast from every predator but man. An average bull stands six feet high at the shoulders, weighs a ton or more and makes the ground shake when he engages in a butting duel with another male. Prairie settlers were so impressed by the animal's strength that they mistakenly believed the bison, rather than the climate, had cleared the trees from the prairie.

Evolution equipped the bison — and many other species of grassland mammals — with another mechanism for self-defense: the tendency to congregate in groups. The bison moved in herds of about 50 to 200, and usually it was only the ill or aging straggler that fell prey to roving packs of wolves. Even several distinct herds stuck fairly close together. When alarmed, the animals stampeded in dark masses that sometimes blanketed dozens of square miles.

Normally the herds moved only when prompted by the need for more forage. Winter posed no serious problem for the great beasts; their thick, shaggy pelts protected them from the cold, and they simply nuzzled aside the snow to get at the well-cured hay beneath. A few acres of tallgrass might sustain one animal for an entire year, but in the shortgrass plains much farther west, something closer to 100 acres per capita was required. The herds did not overgraze an area; they invariably moved on in time to allow the grass to recover.

During the midsummer mating season, male bison established a shifting hierarchy of supremacy among themselves with ferocious, head-butting fights that were awesome to behold. Paired off for combat, two bulls would slam their foreheads together with short, straight lunges or try to drive the tips of their short, curved horns into the other's head and shoulders. Often, at no apparent signal, a general melee would break out. During these so-called fighting storms, as many as 50 or 60 individual battles might take place, arranging and rearranging the social order.

Even more than the urge to mate, the desire to dominate other males motivated the bulls' aggressive behavior. A bull might even abandon a receptive cow to pick a fight with a distant male. Their relationships with one another so preoccupied the bulls that they could think of little else, even eating. As a result, a typical bull might lose 200 pounds, or one tenth of his weight, between June and October.

Few of the fights were fatal, and many challenges succeeded or failed even before the head-knocking began. The bison had developed a set of signals — both vocal and physical — that enabled them to discern one another's intentions. The bellow of a challenging male could be heard for miles across the prairie like a roll of distant thunder. At closer range, the bull telegraphed his fury by snorting, stomping and circling — sometimes turning broadside to his opponent and drawing up his body stiffly to maximum size. Such elaborate preliminaries gave the opponent plenty of opportunity to surrender — either by retreating or by turning his head to the side in a recognizable way. Even after the head-on charge had begun, one bull might signal submission by turning away, and the collision would be avoided by inches. In such cases, the victorious bull almost never pressed his advantage, and both animals lived to fight another day.

Before the bison was hunted to the verge of extinction late in the 19th Century, it shared parts of the prairie with another ruminant, the pronghorn. Often referred to as an antelope, the pronghorn actually combines characteristics of that animal and of the goat — hence its scientific name *Antilocapra americana,* or American goat antelope. Once almost as numerous as the bison, the pronghorn tended to congregate in the drier, shortgrass plains west of the bison's primary range. The two species overlapped in the mixed-grass prairie, but there was little competition for forage. The pronghorn dined mostly on forbs and shrubs that the bison disdained. It even consumed the prickly-pear cactus, a thorny but valuable source of moisture during time of drought.

Lacking the bison's strength, the pronghorn has other assets. Endowed with vision as acute as that of a human equipped with eight-power binoculars, it keeps a sharp watch for enemies. The appearance of a predator a mile or two away triggers the pronghorn's instinctive alarm system. Muscles in the rump contract, raising a patch of white hairs that resembles a big powder puff. This signal — flashing in the sun "like a tin pan," as a pioneer naturalist put it — is repeated throughout the herd.

Thus alerted, the pronghorn demonstrates its other great gifts: the speed and endurance of North America's swiftest mammal. It can bound along for miles at 45 mph and even briefly step up the pace to 60 mph if severely

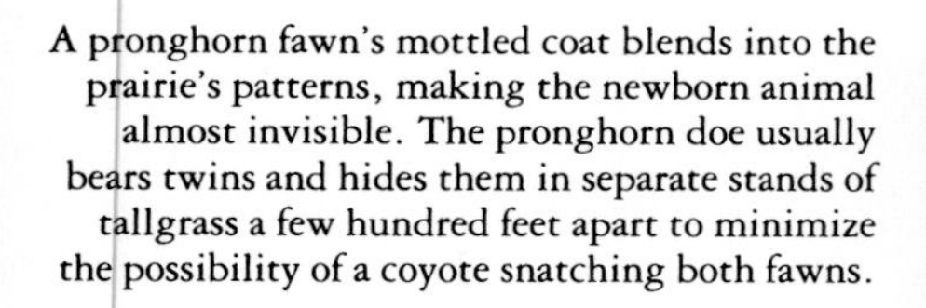

A pronghorn fawn's mottled coat blends into the prairie's patterns, making the newborn animal almost invisible. The pronghorn doe usually bears twins and hides them in separate stands of tallgrass a few hundred feet apart to minimize the possibility of a coyote snatching both fawns.

A female saiga and her foal forage the snowy Eurasian steppe. The saiga's ability to survive the harsh steppe winter is made possible by a brutally effective adaptation: Every year during the rutting season, about four fifths of the males starve themselves to death, leaving most of the limited food supply to the females, which propagate the species.

pressed. Unlike many mammals, it runs in a straight line. This habit enables it to outdistance wolves and coyotes, which often attempt to wear down prey by running in relays.

Superb physical equipment sustains the pronghorn's speed and stamina. It has long legs, a heart twice as large as that of other 100-pound animals and a trachea of extraordinary width. Taking full advantage of this remarkable respiratory apparatus, the pronghorn runs with its mouth open and tongue hanging out, thereby gulping the maximum amount of air.

The pronghorn's counterpart on the steppes of Eurasia is another swift, antelope-like creature, the saiga. Perhaps because it stands only two and a half feet high, the saiga occasionally leaps into the air to scan the horizon for predators. When alarmed, a saiga can reach a speed of 50 mph. But normally it walks along in a curious, shambling gait with its head inclined toward the ground.

This incongruous gait and the configuration of its head have long intrigued biologists. In fact, the two characteristics may be related. The head tapers to a distinctive bulbous snout. Inside each nostril is a large sac lined with a mucous membrane. Some biologists believe this unique anatomical feature serves to filter out the noseful of dust that the saiga inhales as it races, head down, over the dry steppe.

Another adaptation of the saiga, its nomadic habits, helps ensure adequate supplies of water and food. When the forbs and Stipa grasses of the steppe lose their moisture in August, the saiga migrates toward rivers and lakes in enormous aggregations of up to 100,000 animals. Then, in winter, herds move south in search of forage, sometimes traveling 150 miles in a single week.

Even so, saiga herds may get trapped by the early snowstorms that often assault the steppe. As many as 150,000 saigas have perished when ice and snow coat the grass, cutting off their forage. Such calamities, together with the depredations of man, reduced the saiga population, which once numbered in the millions, to only a few hundred about 50 years ago.

But the animal's extraordinary reproductive capacity helped save the saiga. The female saiga is ready to mate at the age of seven months, and about 65 per cent of the time she gives birth to twins — a rarity among hoofed mammals. This adaptation, augmented by the Soviet government's vigorous preservation effort, gradually rebuilt the herds. By the 1970s, the Eurasian steppes harbored about one million saigas.

Despite their size and numbers, the large mammals were seldom the main consumers of grassland vegetation. In many areas of the prairie, little rodents, such as ground squirrels and gophers, vastly outnumbered the bison and pronghorn, consuming in small bites a far greater quantity of forage. It has been estimated that a four-square-mile tract of the present-day Eurasian steppes may support as many as 325,000 rodents and perhaps only four saigas. The four saigas may eat about 40 pounds of forage daily during the spring and summer, while the rodents consume more than two tons of grass, bulbs and other vegetation.

The rodents survive on the open grasslands by evolutionary strategies that differ from those of the big mammals. The little creatures lack both strength and swiftness; only the long-legged North American jack rabbit, with a top speed of 45 mph, can come close to matching strides with the saiga or pronghorn. Therefore the rodents developed a powerful instinct for seeking protection by burrowing underground.

A North American pocket gopher can excavate up to 300 feet of tunnel in just one night. At that rate, it can quickly dig a tunnel system half a mile long and live there safe from fierce blizzards and prowling coyotes.

Among the grassland rodents of three continents, digging and social gregariousness tend to go together. On South America's pampa, beaver-like vizcachas gather in underground colonies of perhaps two dozen individuals. Working together, the vizcachas scoop out cylindrical underground galleries so large and elaborate that just the entrance may be six feet deep and nearly as wide. They dump the soil aboveground in mounds several feet high that give them a commanding view of the flat landscape and an early warning of approaching predators. The vizcachas further enhance their view by clearing the surrounding area of all tallgrasses. Despite their precautions, the vizcachas are so wary of foxes and hawks that they ordinarily stay underground during daylight hours, emerging only after dark to nibble on grasses and forbs.

The vizcacha's counterpart on the Eurasian steppes, a little ground squirrel called the suslik, spends even more time underground. Most species of suslik hibernate for about half of the year, avoiding the worst climatic extremes of the steppe — searing summer winds and the blizzards known as burans, which often rage with hurricane fury for days at a time.

The suslik spends much of the rest of the year preparing for hibernation. It insulates its underground den with dried grass and stashes seeds there, transporting them in its cheek pouches. In some regions the suslik goes underground as early as late June, when the grasses begin to wither for lack of moisture. The creature seals up the burrow entrance and retires to its den, there to fatten up on its store of seeds and then fall into a long sleep, existing on reserves of body fat until the following spring.

Though both the vizcacha and suslik live in groups, neither has evolved with the intense social gregariousness that characterizes the archetypal burrowing rodent of the North American grasslands, the prairie dog. Actually

a member of the ground-squirrel family and no kin to canines, the prairie dog probably got its name from its repertoire of at least 10 different yip-like barking sounds that serve a variety of social purposes — most important, alerting the community to the presence of a rattlesnake, hawk, black-footed ferret or coyote.

The prairie dog is a true town dweller. Before extensive poisoning campaigns carried out by farmers and ranchers decimated them, thousands of subterranean communities dotted the western half of the North American grasslands. In 1900 a veritable megalopolis was reported in Texas: a town that sprawled across 25,000 square miles and housed an estimated 400 million prairie dogs.

The prairie dog's basic social unit is the clan, or coterie. A coterie consists of at least several and possibly more than a dozen prairie dogs —

usually a male, a few females and their offspring. The coterie lives and plays together, engaging in affectionate social activities such as grooming each other and even kissing.

A coterie's home territory usually covers about seven tenths of an acre. Its network of tunnels often contains such amenities as special chambers for use as toilets and extra exits for quick escape when a ferret or badger slips into one of the burrows. Members of a coterie jealously guard their territory, even against intrusion by other prairie dogs, and they pass their burrows down from generation to generation.

An American badger dines on a freshly killed prairie dog dug from an underground lair. Both the carnivorous badger and the plant-eating prairie dog defend themselves against human and other predators with prodigious digging; the badger's burrows sometimes reach nine feet in depth and extend for 1,000 feet under the prairie.

Even while their pup nuzzles them affectionately, a prairie-dog couple remain alert for danger. The adults scan the horizon with high-set eyes well suited for peering over the top of shortgrass and for peeking from the protection of tunnels in which they hide from their enemies.

In their ground-level nest, a hungry brood of horned larks await their parents' return. As adults, these songbirds compensate for the grasslands' lack of trees — where most birds perch to sing — by singing while in flight.

The precise nature of the prairie dogs' social organization benefits them in several ways. The coterie furnishes a friendly environment for successful mating and rearing of the young. Its firm territorial boundaries provide for a uniform distribution of the town's citizenry, minimizing the possibility of overpopulation. At the same time, the grouping of coteries into a larger community serves to protect everyone. When danger threatens, the yiplike alarm signal resounds in a mighty chorus from coterie to coterie, alerting the entire town. At such times, a prairie dog may take temporary refuge in any burrow, regardless of his or her allegiance to a particular coterie.

In addition to the mammals large and small, birds demonstrate their own special adaptations to conditions in the temperate grasslands. Many species differ in behavior from their close woodland relatives. For example, they tend to drink less, to rely more on a seed diet and to flock together more often than their forest cousins.

More than half of the grassland birds build their nests on the ground. They spend much of their time on foot, having evolved strength as walkers rather than fliers. Larks, lacking trees from which to serenade, do most of their singing on the wing, as do the bobolink, the dickcissel and several species of longspurs. Both the Eastern and the Western meadow lark are found in the tallgrass prairie. For some reason, the Western meadow lark sings a longer and more melodic song than its Eastern relative.

Among the birds of the temperate grasslands, some of the most unusual adaptations occur in the ostrich-like rhea of the South American pampa. This earthbound bird, standing five feet tall and weighing up to 50 pounds, fills the biological niche of large hoofed mammals, which did not evolve in the pampa. The rhea compensates for its inability to fly with extraordinary fleetness of foot. With one wing raised like a sail, it strides across the grassy flats at speeds up to 40 mph.

The big birds, once fairly numerous on the Argentine pampa, gradually dwindled. Competition in the form of cattle was introduced at an early date, and the herds multiplied rapidly. By the 1830s, when the great naturalist Charles Darwin visited Argentina, an estimated 80 million head of cattle roamed the plains. The rheas were not only crowded out by the cattle but killed for sport by callous gauchos on horseback, who lassoed the birds with bolas — weighted leather thongs that entangled the rheas' long legs. By 1980, there were only a few thousand rheas left on the pampa.

Of all the creatures that live on the prairie, steppe and pampa, the biggest consumers of grasses are the hordes of insects and invertebrates aboveground and belowground. The largest consumers of grasses are the wormlike nematodes. But the most conspicuous consumers are omnivorous grasshoppers.

Grasshoppers live on all grasslands and are common on the North American prairie. Every prairie state can count at least 100 species within its borders, and Kansas alone has something like 300 species. But only a few species — notably the migratory types commonly called locusts — ever reach such abundance as to fulfill the grasshopper's fearsome reputation.

Even among the locusts, a combination of constraints ordinarily prevents precipitous growth of the population. Birds and other animals prey upon the creature and its eggs; diseases and parasites plague it, especially in relatively humid conditions. The normal loss rate is so high, in fact, that

Displaying his distinctive colors, a yellow-headed blackbird perches among the bulrushes of a North American prairie marsh. Yellow-headed blackbirds consume large quantities of oats, wheat and other cultivated grains, but they help the farmer by keeping beetles, caterpillars and grasshoppers under control.

A male rhea, exploiting its five-foot height to spot natural predators from afar, struts across the Argentine pampa with its young. Human hunters have reduced the rhea's numbers, as have barbed-wire fences, which entangle the flightless bird's legs and block its normal migratory routes.

In a springtime ritual, a greater prairie chicken raises tufts of neck feathers and inflates orange air sacs before emitting its booming mating call. The conversion of North America's natural prairie to farmland provided new sources of food for the prairie chicken, but that gain was more than offset by the elimination of the ground cover in which it nests.

even though a female grasshopper typically lays up to several hundred eggs, only a single pair of adults may survive to breed.

Thus, the dreaded migratory locust is normally just another herbivorous insect, clipping off an occasional stem of grass — actually eating about one half to two thirds of what it cuts down — and making scarcely a dent in the prairie.

From time to time, however, environmental events conspire to send the locust population spiraling out of control. Prolonged warmth in autumn may extend the egg-laying season, and a cool spring may delay the hatch until the grasses have grown enough to provide ample food for the young nymphs. Meanwhile, the normal constraints break down. Successive years of drought inhibit the diseases that curtail the locusts. Many birds and other predators migrate elsewhere in time of drought and thus fail to consume their usual insect quota.

The locust population multiplies geometrically. Crowded together in numbers that may approach an incredible density of 1,000 insects per square yard, the creatures become a dangerous force. They rise up in gigantic swarms like the plagues of locusts described in the Bible.

So it was in the prairie during the 1870s. The species *Melanoplus spretus,* commonly known as the Rocky Mountain locust, arose multitudinously from its customary habitat in the dry shortgrass plains at the foot of the Rockies and, riding the prevailing westerly winds on long wings, swarmed eastward.

The novelist O. E. Rölvaag was a horrified witness to the locust invasion of South Dakota. He wrote that the hordes of insects descended "with terrific speed, breaking now and then like a huge surf, and with the deep dull roaring sound as of a heavy undertow rolling into caverns on a mountain side. It seemed as if the unseen hand of a giant were shaking an immense tablecloth of iridescent colours."

In Nebraska, other observers roughly established the dimensions of the invasion of that state. The locust swarm, moving eastward at about five mph, took six full hours to pass one observation point. This meant the great mass was about 30 miles deep. It was advancing on a front at least 100 miles wide and nearly a mile high. By these estimates, the invasion consisted of 124 billion locusts. The insects devoured everything green in their path. They swept the fields and pastures cleaner than prairie fires do.

When the ravaging hordes reached the more humid tallgrass country, the threat slowly subsided. Within a year after the deepest eastern penetrations, the grasses, their intricate root systems still intact, gradually recovered. But not the Rocky Mountain locust. During the next half century, the insect inexplicably disappeared.

The prairie's ability to withstand locusts as well as fire and drought demonstrated the dynamic balance of all grasslands in their natural state. But the resilience of the primeval prairie has long since been reduced by intensive agriculture and heavy grazing by domestic herds, which strip away the grasses and expose the earth to the forces of erosion. More than three decades ago, ecologist John Weaver realized that irremediable damage had already been done and wrote a lament for his beloved grasslands. "Prairie is much more than land covered with grass," he said. "It is a slowly evolved, highly complex organic entity, centuries old. It approaches the eternal. Once destroyed, it can never be replaced by man." **Ω**

THE LIVING SOIL

The inert appearance of grassland soil is deceptive. Actually it is a dynamic environment that teems with life. Ever changing, the soil erodes and is replenished by the gradual weathering of rock into clay, silt and sand. It is also enriched by the activity of myriad underground plants and animals.

The soil-building process is exceedingly slow: The surface layer erodes naturally at a rate of no more than one inch each century. At the same time, the rocks beneath the surface are breaking down into smaller and smaller pieces; eventually these new soil particles come within the reach of grass roots.

Grasses are both the beneficiaries of the soil's largess and contributors to its regeneration. An average of 70 per cent of plant tissue in grasslands lies underground, where intricate root systems absorb nutrients as well as the water and air that filter down from the surface. In return, the plants anchor the soil against punishing weather and fertilize it with their organic matter when they die.

Once a blade of grass falls to the ground, it is attacked by an army of microorganisms and invertebrates. They are part of an intricate food chain that in most grassland areas includes more than 1,000 species that feed on both dead and living plants and animals.

The end product of this complex food system is the lush, dark brown substance called humus. Made up of partially decomposed plant and animal matter, humus forms the organic portion of grassland soil. It is so absorbent that it can increase by 20 per cent the soil's capacity for holding water, and so cohesive that it binds mineral particles into the clumps characteristic of fertile soil.

Humus also provides nitrogen and other elements vital to plant growth that are absent from the soil's mineral particles. In fact, though humus accounts for at most only 10 per cent of the grassland topsoil, without it the soil would be as barren as the surface of the moon.

A seedling lupine, a wild flower of the American Great Plains, nudges aside clumps of nurturing soil as it emerges into the sun. Dried stems and roots of earlier plants, visible in the dirt, will decompose and enrich the soil in a continuing cycle of renewal.

As this tiny rivulet freezes, it will widen the channel it has eroded through a rock. The expansion of the water as it turns to ice exerts 150 tons of pressure per square foot on the rock's surface, wrenching loose tiny mineral particles; come spring, the thawing water will carry the particles to the ground below to become new soil.

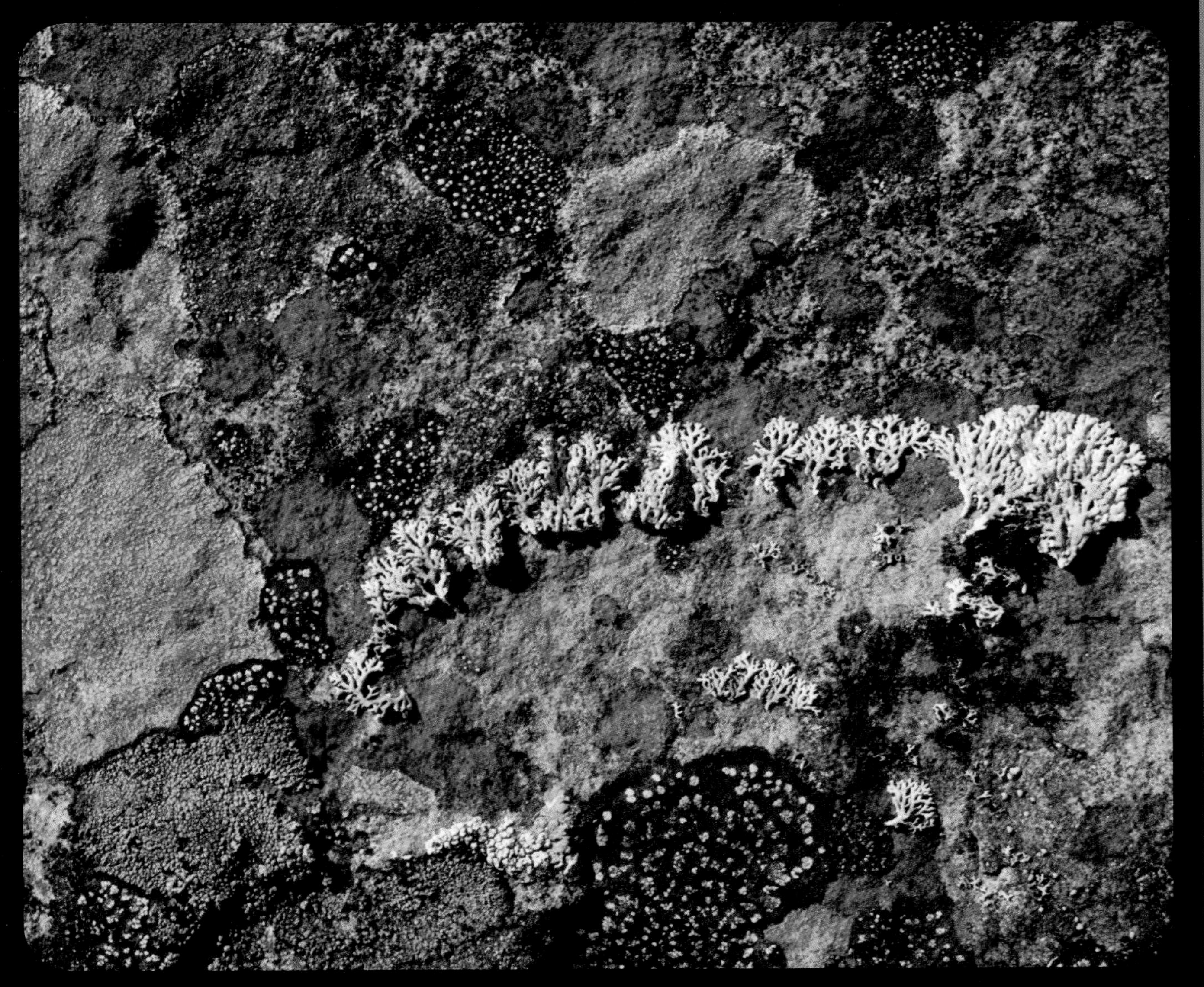

Patches of lichen in varied colors coat a tundra rock, their sprightly appearance masking their corrosive power. Each lichen chemically removes micas, feldspars and other susceptible minerals from the surface of the rock to which it clings, creating a dusty soil in which plants of a higher order can flourish.

Underground Pathways for Air and Water

Abundant humus darkens the topsoil layer, called the A horizon, in this soil profile. Percolating water leaches minerals from the A horizon into the lighter, crumbly B horizon *(center)* or the still-lighter C horizon *(bottom),* which is streaked by dried pathways of percolating water.

Glistening like diamonds, crystalline grains of sand mingle with minute particles of clay *(center)* and intermediate-sized silt *(bottom)* in this sample showing the three particle groups of which soil is composed. The larger sand particles enhance drainage and air movement, while the smaller particles readily form clumps that retain water and water-soluble elements.

Cavernous pores separate clumps of soil particles in this close-up view. Ideally, half of the volume of fertile topsoil is made up of such pores, which hold the air and water that are requisite to plant growth.

The Beneficial Exchange of Nutrients

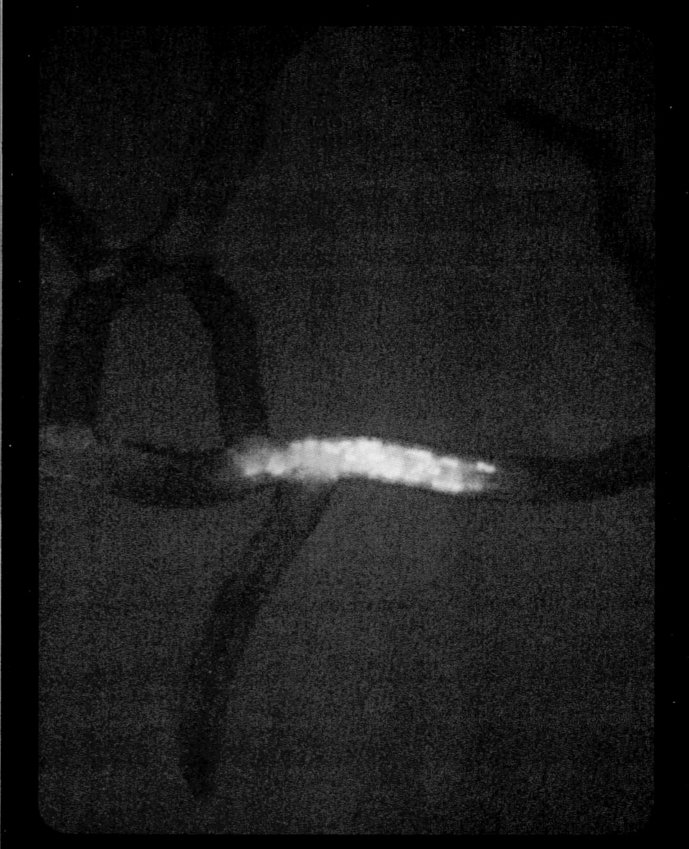

Portions of a microscopic fungus fluoresce within channel-like grass roots when illuminated by ultraviolet light. In a mutually beneficial relationship, the fungus draws simple sugars from the grass and in exchange provides the plant with vital nutrients.

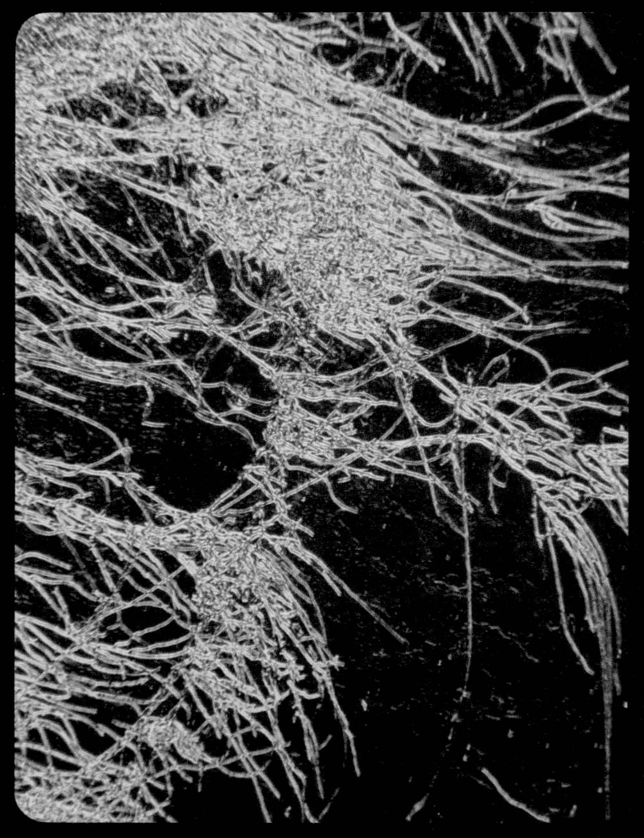

Blue-green algae collect on the prairie mud after a rain. Flourishing on tiny beads of water at or just below the surface of the soil, these aquatic organisms convert nitrogen in the atmosphere into compounds that can be absorbed through the soil by grassland plants.

A tiny mushroom, dwarfed by the prairie grass around it, is one of 700 species of fungus that thrive in grassland soil. The fungi extract sugar and other substances from plant tissue, thus contributing to the decomposition process that produces humus, the soil's fertile component.

Armies of Insects That Enrich the Land

A red velvet mite scurries through clumps of soil underground. Using the narrow pores in the soil as their roadways, the mites transport decomposing matter below the surface, thus deepening the layer of fertile topsoil.

Tiny insects known as springtails *(left)* are among the first organisms to attack dead plant debris; they chew holes that expose plant tissue to the bacteria and fungi that decompose it. The springtails in turn are preyed upon by beetles and other invertebrates.

The translucent white potworm, like its relative the earthworm, eats its way through soil, ingesting both minerals and decomposing organic matter. Its crumblike secretions, or casts, help stabilize the soil and increase its fertility.

Minute nematode worms such as the one shown at right abound in the upper layer of moist soil, clustering around the plant roots from which many parasitic species of nematode draw nutrients. The nematodes are themselves consumed by mites and other predators.

A Profusion of Roots for Absorbing the Soil's Largess

Tiny hairs sprout from the roots of Western wheatgrass, expanding the area from which the grass draws air, water and nutrients. This area, called the rhizosphere, is particularly rich in microorganisms whose growth is stimulated by the roots' organic secretions.

The complex root system of a Jack-in-the pulpit harbors a single earthworm *(center)* as th plant is pulled from its bed. The five-inch long worm bores through the upper soil, leavin holes that help to aerate and drain i

A tufted evening primrose clings to life in the parched grassland soil, its roots insulated temporarily from drought by a layer of gelatinous secretions. This coating maintains contact with the nurturing soil even as the drying roots of the primrose shrink.

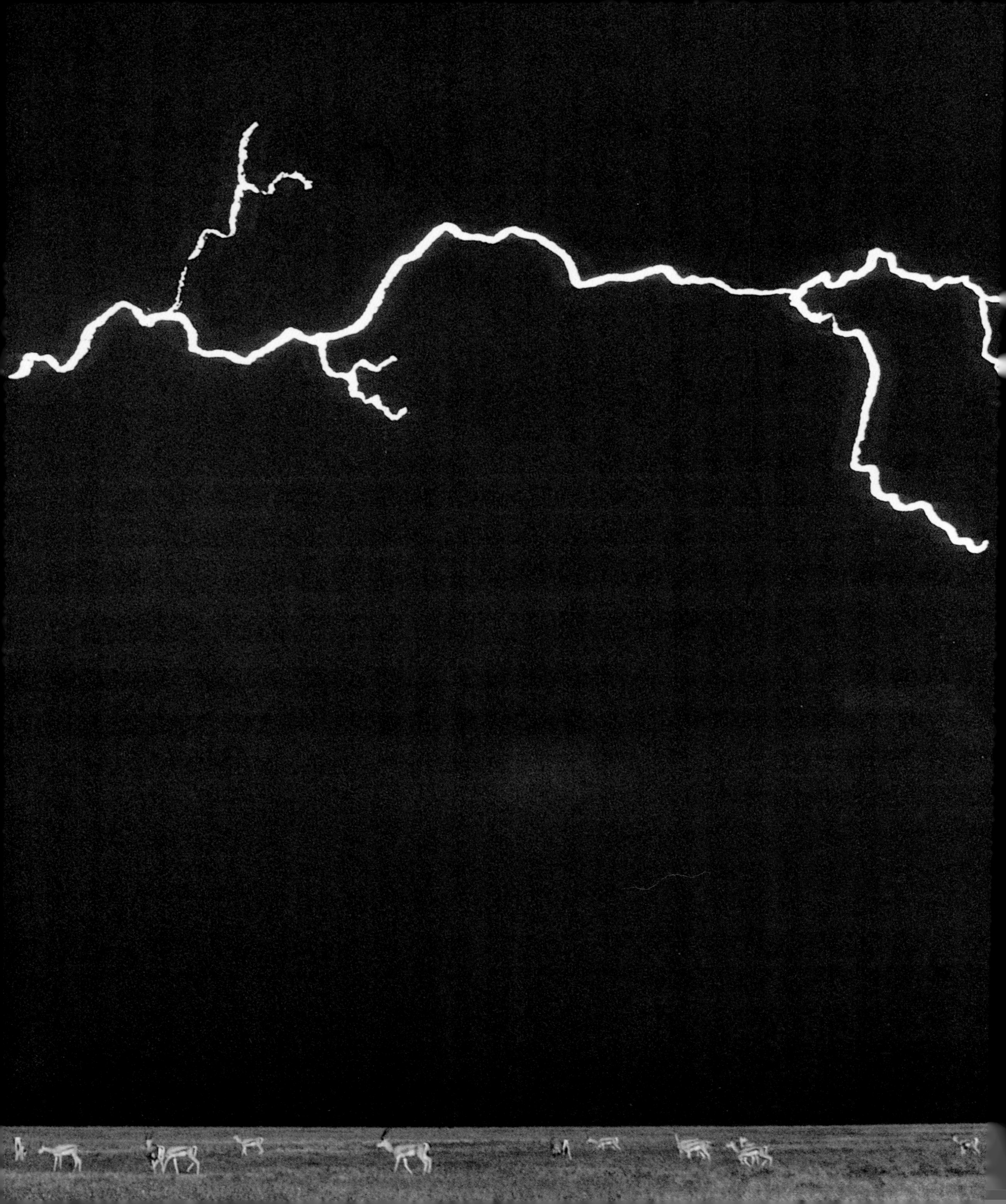

Chapter 3

THE GLOBE-GIRDLING SAVANNAS

When the rains finally come to the great Serengeti Plain of East Africa, it is an overwhelming event, almost theatrical in its drama. For months the air has been thick with dust; the tall grasses have become dry and brown, and the branches of the gnarled acacia trees that dot the plain are brittle from lack of moisture. In many places, scorching fires, set by herders to clear away dead growth, have blackened the ground and made the burnt grass as sharp as glass. Then one day, almost imperceptible flickers of lightning appear in the distance; the temperature rises, and the atmosphere becomes unbearably oppressive. Clouds begin to build, then disperse; but a sense of anticipation fills the air. At length there is a blast of cool wind, the sky darkens rapidly and thunderclouds explode, releasing sheets of rain that soak the earth and transform dry riverbeds into raging torrents.

Almost overnight, the Serengeti comes exuberantly to life. New grass shoots up so fast — as much as an inch in 24 hours — that onlookers have the bizarre sensation that they are seeing it grow before their eyes. Within days, the grass is six inches high — and it will eventually reach heights of from one to six feet. The plain is now a dazzling shade of green. The acacia trees are festooned with yellow blossoms, the air is clear and exhilarating, and the wild grasses sway as far as the eye can see. The land, so recently empty and inhospitable, suddenly abounds with animals. The great migrating herds have returned — zebras, wildebeests and Thomson's gazelles by the hundreds of thousands, newly arrived to devour the fresh foliage.

The herds will remain as long as the grass keeps yielding vigorous new growth — throughout the wet season, which in the Serengeti lasts from five to six months. When the ground again dries out and the grasses stiffen and become dormant, the massive herds will abandon the area and move on, migrating as they have done for thousands of years in the unending quest for food. They share the Serengeti with a wondrous range of nonmigratory species: lumbering elephants; dainty dik-diks; towering giraffes; high-bounding impalas; families of ostriches; as well as African buffalo, baboons, lions, leopards, rhinoceroses, waterbuck and warthogs. Together they form a constantly shifting kaleidoscope of grazers, browsers and predators, augmented by flocks of multicolored birds, ubiquitous insects and a population of small animals and reptiles that live close to the soil or beneath it. Few other environments on earth support such a diversity of life.

The Serengeti Plain is a savanna, the most prevalent of the world's grasslands. Savannas cover six million square miles of the earth's surface, ranging for the most part in broad tracts on either side of the Equator in latitudes

Lightning crackles above a herd of gazelles on Africa's Serengeti Plain, heralding the end of the long dry season. The oncoming rain will transform the parched savanna into a verdant Eden.

where the weather is always warm, with daytime temperatures rarely dipping below 70° F. These tropical and subtropical plains are characterized by distinct wet and dry seasons of varying length — the months of dramatic, life-giving downpours alternating with extended periods of drought.

Grass is continuous and dominant in the savannas, outproducing all other forms of plant life. But in most places savannas also support a scattering of trees and certain other woody plants that can withstand the long dry seasons. One other element defines savannas: fire. Every year during the dry season, fires burn off wide areas of the plains. These blazes play an integral role in the maintenance and enlargement of the savanna.

Extensive areas of savanna exist in South America — the llanos, or flatlands, north of the Amazon rain forest, and the campos, or tablelands, south of the rain forest. Large stretches of Australia are savanna, and there are patches, too, in Central America, India and Southeast Asia. But it is in Africa that the savanna predominates, sweeping across fully one third of the continent in vast, parklike plains and gently rolling foothills. The African savanna stretches from the Atlantic Ocean eastward to the Indian Ocean in a solid belt 3,000 miles long, from Guinea to the southern Sudan, Uganda and Kenya. Then it reaches southward and westward across Tanzania and southern Zaire to Angola. It is in protected areas of Africa, such as the Serengeti National Park in Tanzania, Kenya's Nairobi National Park and the Kabelega Falls Park of Uganda, that savannas survive in something close to their primeval state.

Scientists have been slow to agree on how the savannas originated. There is a consensus, however, that climate, soil and fire have each made an essential contribution to their formation. Enormous areas of savanna, for example, owe their origin to the equatorial climate cycle in which prolonged spells of withering drought are followed by equally extensive periods of rainfall that fluctuates in different locations from 20 to 60 inches annually. Savanna grass is uniquely equipped to deal with both conditions. When the parts of the grass aboveground die out during the dry months, the belowground parts, a rugged mass of fibrous roots that often penetrates the soil to depths greater than the peak height of the grass aboveground, remain alive and well, storing starch reserves and soaking up any available moisture.

Even during the dry season, the difference between day and night temperatures near the Equator is so dramatic — as much as 40° F. — that it causes condensation belowground; each grass rootlet is encased in sand and can absorb the moisture lodged between the grains. When the rains come, grass, which grows from the base of its leaf, is able to respond with much greater speed than shrubs or trees with growth buds farther from the ground. Within an astonishingly short time after the first rain, the vigorous bunch grasses that cover much of the savanna are producing from 5 to 18 tons of green growth per acre. These grasses are not only well adapted to the seasonal cycles but actually seem to require them to maintain their resilience. If the rainfall were spread out evenly over the whole year, large sections of savanna might well turn into woodland or dense tangles of bush.

Some types of savanna, however, cannot be explained by climatic conditions. There are high-rainfall areas in parts of Zaire and Uganda, for example, where sizable stretches of savanna occur suddenly in the midst of thick forest; it is possible for a traveler to be surrounded by dense trees and,

Clumps of tough spinifex, or porcupine grass, thrive in a desiccated region of the Australian interior. Spinifex grasses tolerate the high salt content usually found in dry soil and thus grow where few other plants can.

seconds later, to be out in the blazing sunlight of open grassland. Such a radical shift within such a short distance cannot possibly result from seasonal wet and dry cycles. These areas, known as edaphic savannas, are caused by soil conditions. They occur on hillsides and ridges, where the soil is too shallow and the underlying rocks are too close to the surface to permit large trees to root. Even if the rainfall occurred year round, these edaphic savannas would not turn into heavy forest.

Fire is the third factor in the formation and preservation of a savanna, and it may well be the most decisive element of all. Natural fires, touched off by lightning, sweep across vast areas of dry grassland every year. But man is the great fire maker, and to him goes responsibility for expanding the earth's savanna land to its present boundaries. Whole sweeps of savanna have emerged from the ashes of forest fires started by man to open up grazing territory. And year after year, the spectacular conflagrations that light up the savanna night have been set by farmers and herders to clear away old growth, control the spread of trees and get rid of predators. (Contrary to legend, however, large carnivores do not flee in panic from fire; lions in fact sometimes go extremely close to the flames and often lie down in the warm ashes.)

If the fires stopped for three or four years and if soil and climatic conditions were favorable, the woody seedlings that almost always are intermixed with savanna grasses would shoot up in great numbers and thrust out branches high enough to escape the fires when they eventually returned. As it is, some seedlings manage to survive the fires, and many savannas, particularly those in Africa, are dotted with trees. These hardy specimens possess either thick, corky bark or smooth, resinous bark that makes them virtually fireproof as well as drought resistant. Some, like the colossal baobabs, which are capable of storing up to 25,000 gallons of wa-

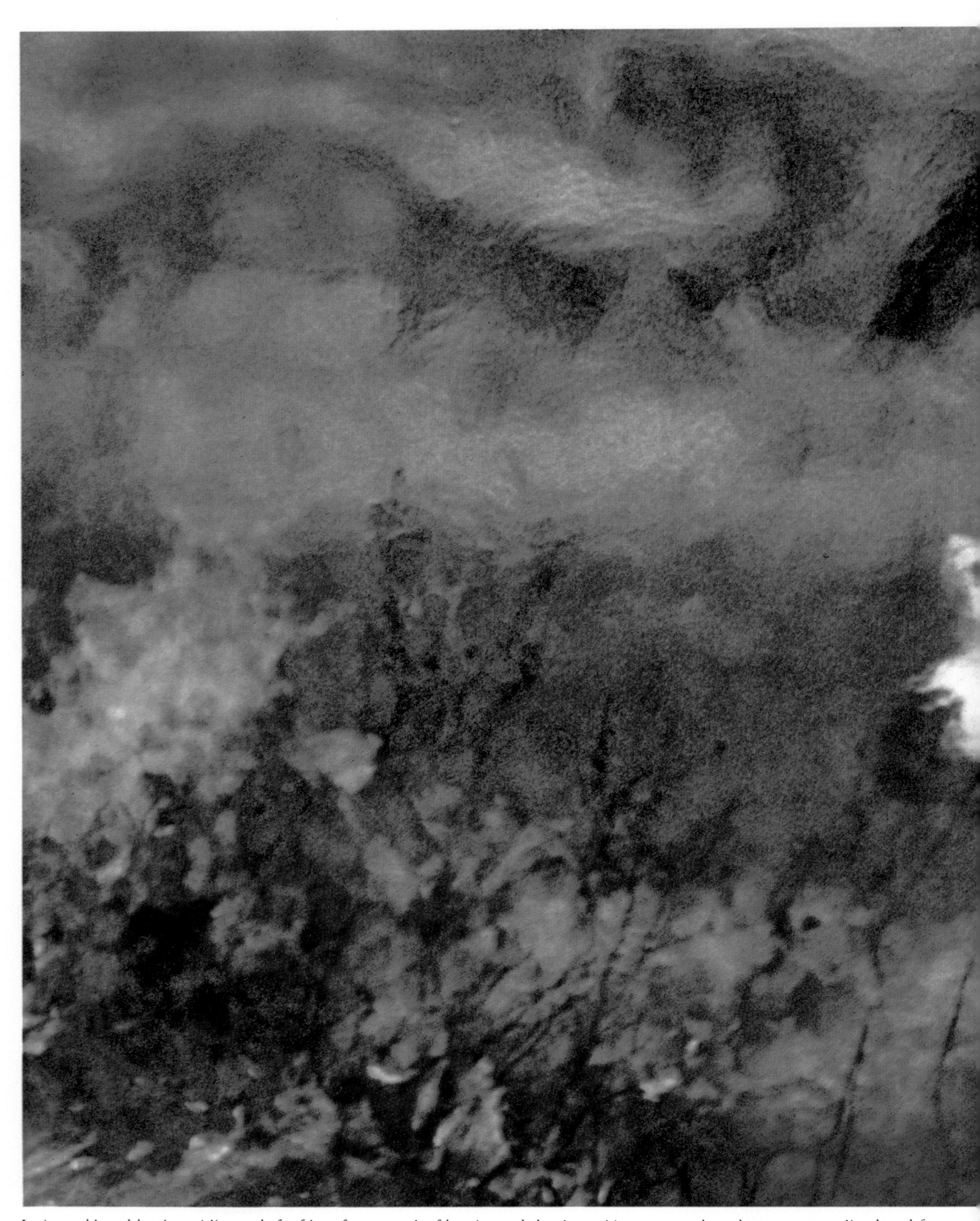

Its image blurred by the swirling updraft of heat from a patch of burning underbrush, a white crane stands ready to pounce on lizards and frogs as t

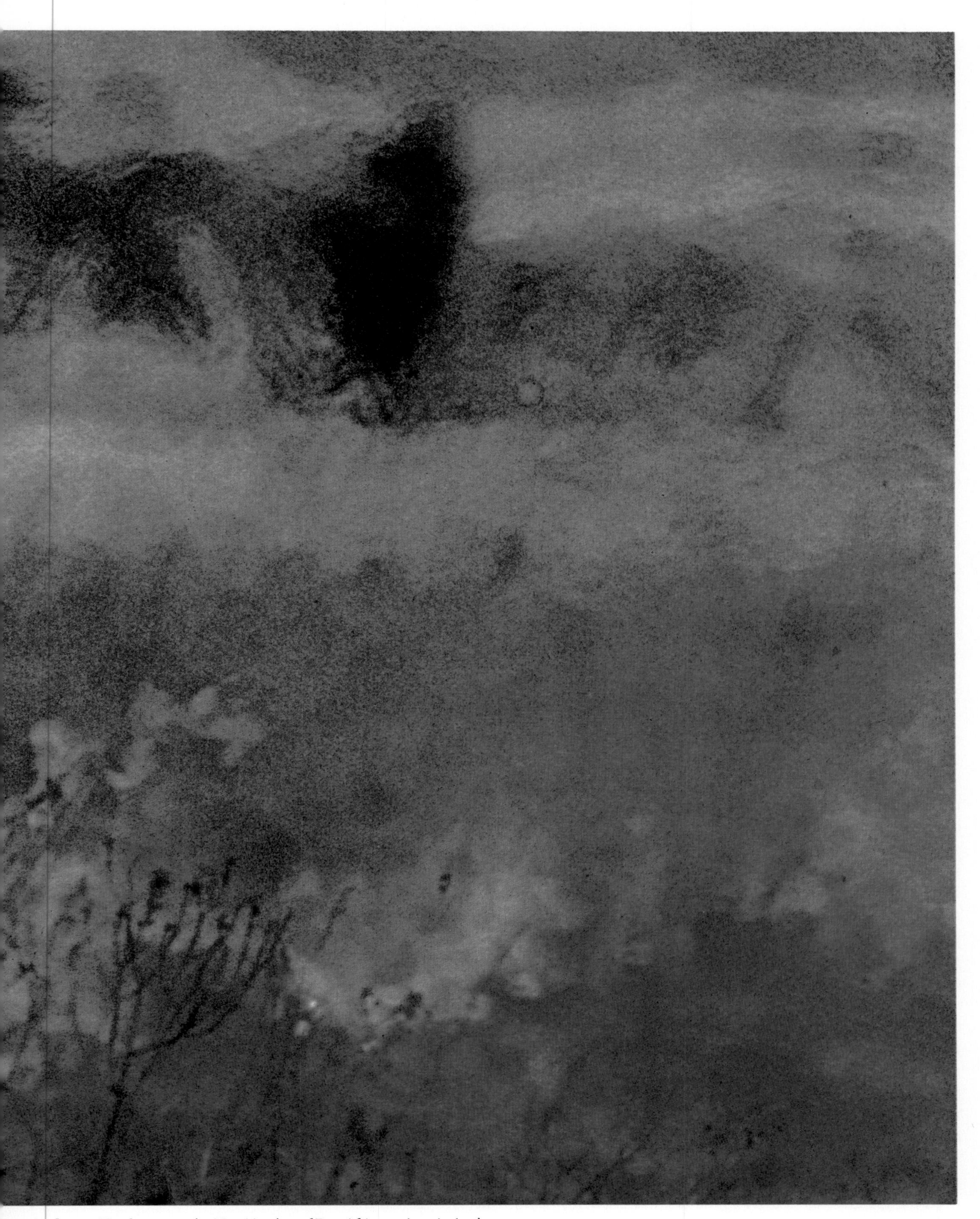

ape the flames. The fire was set by Masai herders of East Africa to clear the land.

ter, attain an age of nearly 100 years; but most, including the lofty, feathery acacias, whose foliage is favored by giraffes, live only a few decades—rarely more than 60 years.

While it is not known with any degree of certainty when savannas first appeared, the large herbivores that populate them clearly evolved either in response to the new grasslands or in conjunction with them in a form of co-evolution. Ancient fossil remains dug up by paleontologists reveal in fascinating detail the evolutionary changes in animals that survived successfully over the ages in the savanna. The small number of toes on hoofed mammals, for example, reflect the advantages of being able to run swiftly to escape predators in open country and to travel great distances in search of food because of the alternating wet and dry seasons. Another adaptation is reflected in the teeth of most grazing animals: The thick, ridged crowns protect the teeth from the destructive wear and tear of chewing grass close to the ground and picking up abrasive sand and dirt in the process. The wide lips of hippopotamuses are ideally adapted to mowing short swards of grass when the hippos emerge from the water at night to feed.

Until it was hunted almost to extinction, the Orinoco crocodile ruled the swampy llanos of Venezuela. Despite its fearsome appearance, the carnivorous beast is a relatively light eater: One feeding can last it a week. The crocodile also swallows pebbles and stones as ballast so that it can float right side up.

At first sight, the typical savanna may appear to be a flat plain, its covering of wild grass interrupted only by a few baobab and acacia trees. Actually the savanna is a combination of lowlands and highlands, broken by ravines, rivers and intermittent streams. Even the seemingly level plains are a series of undulations ranging from a few feet to hundreds of feet in height. Distances varying from several feet to several miles separate the hillcrests. These undulations determine the distribution of the savanna's tallgrasses and shortgrasses; the grass at the bottom of depressions grows longer because of the water that drains from the top of the hill.

In addition, what appears to be a monolithic spread of one species of grass turns out on closer scrutiny to be a profusion of diverse growth. Savanna grasses generally are taller and coarser than temperate grasses and grow in tufts and bunches rather than as uniform ground cover. The bunches are usually separated by smaller perennial grasses that make remarkably proficient use of sunlight, moisture and the available nutrients in the soil. The species are amazingly numerous—55 kinds of guinea grass and 43 of dropseed in Tanzania alone. Hood grass and bluestem grass are common everywhere, and red oat is probably the most abundant and best known. Soil and topography determine which species will dominate which areas. Rhodes grass and red oat, for example, flourish in well-drained soil, where hood and guinea grasses are weak. Bristle grasses dominate where the soil is claylike, and very tall grasses, such as the impenetrable elephant grasses, prevail where the soil turns boggy. Wiry grasses, which thrive on highly alkaline soil, cover the shorelines of shallow lakes. The tall perennial Napier grass is generally found only in places where the forest has been burned to create new savanna—and continues to need fire to maintain its supremacy. The sweetest grass of all, star grass, occurs mostly on the top of termite mounds—which can be huge—where termites have created edifices full of elaborate tunnels that aerate and drain the soil.

Nearly all savanna grasses are palatable, and at least when young they make good fodder. At the start of the rainy season, the first shoot appears near the ground; the entire plant is then growing tissue with thin cell walls that animals can eat without destroying the growing point. Later the stem

Propelling themselves through water with small, webbed paws, capybaras are well-suited to South America's wettest savannas. Though they are the world's largest rodents, weighing 100 pounds or more, capybaras lead a peaceable group existence, feeding on vegetation found along riverbanks and in the water.

raises the leaves higher above the ground in the competition for sunlight, and the protein content of the maturing leaves is reduced. Finally, the stem extends to form the flowering culm, which contains the lowest food value. The grazing of wild animals and the well-being of savannas are inextricably intertwined. As long as the animals are eating, the grass continues to grow, sending up new shoots. If there are not enough animals, the grass matures into full flower, becoming far less palatable.

The savannas of Africa, South America and Australia share patterns of climate, soil and topography that result in similar ecosystems in which plants and animals interact in like manner. Yet each continent has evolved an independent hierarchy of animal life. South America, for example, lacks the large herbivores typical of Africa, but because vast portions of its savannas are centered on river basins that flood each year, a fascinating variety of aquatic species abounds there.

The rains come in April to the Orinoco River basin of Venezuela and Guyana, north of the Amazon rain forest. Until their arrival, the llanos are arid and baked by the tropical sun. Fires sweep through the region, scorching the grasses but having little effect on the groves of fire-resistant llano palm. Then the rains begin. By the end of the month, the area has been transformed into a flood plain laced with lakes, lagoons and marshes that will remain until the dry season returns in October.

The constantly changing water levels present no problem to the capybaras, the world's largest rodents, which graze both on grasses and on aquatic herbs. Often four feet in length, the stub-tailed capybaras are noisy, social animals. They live in large groups and like to bask on riverbanks so that they can plunge into the water when such land predators as jaguars, cougars and boa constrictors threaten their repose. Once in the water, however, the capybaras are prey to the Orinoco's voracious crocodiles and 20-foot-long anacondas.

Aside from herds of domestic cattle and sheep, the only hoofed animals on the llanos are two species of deer, the whitetail and the red brocket, excellent swimmers that also range into the forests bordering the savanna. South American grasslands have also produced two unique animal species, the giant anteater and the giant armadillo. The sharp claws, tubular snout and long tongue of the six-foot-long anteater enable it not only to attack the large anthills that dot the savanna but to break into the cement-hard nests of termites as well. Thus the anteater fills the same ecological niche that the aardvark does in the African savanna, though the two animals are not related. The 125-pound armadillo has even stronger foreclaws, which it uses for slashing termite hills and for burrowing a safe underground home. The armadillo is further protected by plates of shieldlike armor — the heritage of some ancient ancestor. When danger threatens, it may roll itself into an almost impenetrable armored ball.

A young giant anteater rides piggyback atop its bushy-tailed mother on a foray for food in their native South American savanna. The giant anteaters have no teeth, but with their long, sticky tongues, they can lap up as many as 30,000 ants and termites a day.

South of the rain forest and east of the Andes Mountains, areas of Brazil, Paraguay, Uruguay, Argentina and Bolivia are covered with savannas. Their character varies widely, depending on the seasonal rainfall. Some of the campos of Brazil remain underwater much of the year; other, drier areas are treeless or at most are spotted with thorny shrubs and patches of stunted trees. In Argentina, parts of the campos become hot semideserts where the winds roil the dust into clouds and only the cactus plants, 20 to 30 feet tall, seem able to survive the dry months. Yet even here, the grasses return with the rains, and in summer the region becomes a vast, mosquito-infested swamp, flooded by rainfall and by the melting snows of the Andes.

A number of llanos animals — among them deer, capybaras, anteaters and armadillos — have also adapted to these southerly savannas, along with a maned wolf, which evolved long, slender legs that enable it to sprint at high speeds in its nightly hunt for rodents and ground-dwelling birds.

Rainfall, or the lack of it, also is a principal factor in shaping the savannas of Australia, although fires have played a major part in clearing and controlling the vegetation ever since man first arrived on the scene 32,000 years ago. Australia is the world's driest continent: One third of the land, or one million square miles, receives an average of only 14 inches of rain per year; and another one million square miles of interior desert gets 10 inches or less. Bounding the central desert are diverse regions defined as savanna. They range from lush tropical savanna, north of the Tropic of Capricorn, through parklike woodlands and scrub country, to broad expanses where only grasses grow.

Two centuries ago, early European settlers in Australia found the interior vegetation dominated in many places by tall eucalyptus trees. The eucalyptus required little water — its roots reached deep into the soil for nutrients and moisture — and its canopy of branches was thin enough to enable

sunlight to flood the forest floor, where a thick cover of grasses grew.

Most of these eucalyptus groves have been destroyed by grazing and cultivation. There remain, however, broad stretches of scrubland studded with dwarf eucalypts known as mallee scrub, and a variety of the acacia tree called mulga. The periods of drought are prolonged in the scrubland, and the rainy seasons are brief and undependable. Yet even here tough, nutritious grasses thrive, and when it does rain, flowers appear. For a time the scrubland becomes a magnificent quilt of violet, yellow, pink and white.

Isolated for ages from other continents, Australia developed its own species of browsing and grazing animals. Chief among them are the marsupials, a wide-ranging order of mammals in which the females carry their young in a stomach pouch. Largest of the marsupials are the kangaroos, plant-eaters who use their powerful hind legs to bound across the grasslands at speeds of up to 40 miles per hour. Australia's millions of kangaroos come in many species; some are no more than a foot high, but others, such as the red kangaroo that populates the open, scrub savanna, reach a height of seven feet. Kangaroos have two characteristics essential to survival in Australia's interior: They grow fat on the relatively poor fodder of even the driest grasslands, and if they can find shade, they can do without water for extended periods.

Large numbers of nomadic birds have adapted to Australia's lightly wooded savannas; among them are colorful flocks of parrots, parakeets, cockatoos and the continent's largest bird of prey, the wedge-tailed eagle. But Australia's archetypal grassland bird is the bustard, which lives almost entirely on the ground. The bustard builds its nest in the grass and, even when threatened by such enemies as foxes and dogs, prefers to run away rather than fly.

The ecological balance of Australia's isolated savanna has often been disrupted by alien animal species introduced from overseas. The earliest of these exotics probably was the dingo, or wild dog, which came from Asia at least 5,000 years ago. The dingo preyed with devastating effect on the

A noxious, violet-colored weed known as Patterson's curse spreads through a savanna in New South Wales. Named for a 19th Century rancher whose land it first threatened, Patterson's curse invades overgrazed pasture, spreads rapidly and contains alkaloids that can be lethal to animals if the plant is eaten in large quantities, unmixed with other grasses.

bustard and the smaller marsupials until it too became a target — of man, who correctly regarded the dingo as a mortal threat to his flocks of sheep.

At least 17 other species of wild mammals have been introduced to Australia since European settlers arrived in the late 18th Century. The most destructive was the European hare, which multiplied astonishingly and wiped out several species of grassland plants with its rampant grazing.

Humans, however, have been the most disturbing influence on the Australian savanna. Relentless hunting has driven the once-abundant bustard to find refuge in remote corners of the land. Vast numbers of parakeets once were trapped for the pet-store market and thousands of wedge-tailed eagles were slaughtered each year on the pretext that the eagles preyed on lambs. (In fact, the eagles' primary prey are the rabbits, whose potentially explosive population the eagles control.)

More recently, the main target has been the kangaroo. Long condemned as competitors for grazing land, kangaroos have been destroyed at the rate of a million or more each year. Only lately has it been demonstrated that kangaroos — like most savanna animals — live in balance with their habitat. Instead, the chief culprits in the deterioration of Australia's grasslands are the farmers and herders themselves. They often set fires at the worst possible time of year, when the nutritious plants are carrying seed, and they stubbornly concentrate their flocks in the same areas year after year, so that the land becomes overgrazed.

Between drinks, a flock of shell parakeets swirl above their shallow watering hole to confuse any predators lurking nearby. The parakeets feed on grass seeds, and they follow their food and water supply across the Australian savanna — north in autumn and south in spring.

Only where the savanna is protected is it truly healthy. No better example exists of a thriving savanna ecosystem at work than the Serengeti National Park in Tanzania, perhaps the most important wildlife sanctuary left on earth. For thousands of years, much of East Africa was an Eden, teeming with uncounted millions of wild animals that had evolved in splendid harmony with the environment. Less than a century ago, it was still possible for an adventurous traveler to see the Serengeti in its primitive, unmolested state. Getting there required a tortuous trek inland from the coast and involved thrashing for days through dense and suffocating thornbushes. Then, in the early 1900s the building of the Kenya-Uganda railroad made Eden suddenly more accessible — and nearly destroyed it. Hordes of hunters — both trophy-seeking amateurs and professional hunters lusting after ivory, ostrich feathers, rhino horns and other profitable bounty — invaded the East African savanna. (But not without certain casualties; chronicles of the day tell of two lions that held up construction of the railroad for months by attacking and eating 28 African workmen and the rifle-toting inspector who was supposed to be protecting the men.)

During World War I and again in World War II, the wild-animal population fell victim to wholesale slaughter. Soldiers on both sides killed thousands of the animals for sport and for food. At the same time, efforts to convert large tracts of the savanna into pasture for domestic livestock and into fields for agriculture introduced a new threat to the ecology: shortsighted methods of land management and cultivation.

To halt these depredations, a portion of what is now the Serengeti National Park was set aside as a protected reserve as early as 1929. Today, though gravel roads and airstrips have begun to intrude on the primitive landscape and some poaching takes place, the preserve is as close to natural savanna as 20th Century man is likely to find. More than one and a half million animals still live there, and the seasonal migrations of the great herds remain an awesome sight. The Serengeti in addition contains a variety of habitats, each with a somewhat different animal population; together they afford an illuminating look at the way vegetation of the savanna interrelates with the creatures who live there.

Located along Tanzania's border with Kenya, the Serengeti is bounded on the west by Lake Victoria, Africa's largest lake, and on the east by the Rift Valley. The park itself occupies an area of 5,000 square miles of grassland and open woodland — an area roughly the size of the state of Connecticut — and forms the central part of a 9,000-square-mile area known to scientists as the Serengeti Ecological Unit. Intense studies of the savanna have been under way here for a generation.

As recently as 30 years ago, almost nothing was known about the soil, grass or other vegetation in the Serengeti, and no one had any idea of how many animals were there, what their migratory paths were, why they moved off or where they went. Then Bernhard Grzimek, a renowned German naturalist, was inspired by the ambitious notion of taking an inventory of the area and all the creatures in it. In 1958 he set out across the Serengeti with his 23-year-old son, Michael. "We intended to make a census of this veritable ant heap," Grzimek said, "and to plot the movements of these huge armies of animals."

Traveling in a light plane painted with black and white stripes that gave it the appearance of a flying zebra and in a similarly decorated land vehicle, the two Grzimeks spent two arduous years patiently counting and recounting the animals and tracking their movements. They took thousands of pages of notes and shot endless feet of film. During the growing season they flew as close to the ground as possible — altitudes of 30 to 60 feet — at speeds as slow as 35 mph and landed every time the grass changed color. After each landing they inspected a circular area of about 300 yards' radius, noting which grasses were the most abundant and which were the most commonly eaten. They stuck samples of plants on hundreds of pieces of

Kangaroo grass, a tufted perennial that grows three feet high, is one of Australia's most prolific grasses. During the pollination season, tubular, bright orange anthers and bristly magenta stigmas emerge from the plant's spikelets.

paper and filled dozens of linen bags with soil samples. "All the lists of hundreds of Latin names of plants and grasses clearly showed one thing," reported Grzimek. "The herds migrated to areas where their favorite food was available. The only plants growing in the places that the wildebeests and zebras avoided had little food value, and the animals refused to touch them." After attaching collars to wildebeests and gazelles — and at one point color-marking zebras yellow — the Grzimeks followed the herds when they moved. The naturalists discovered not only that the animals' routes differed from all previous speculation but that they covered far greater distances than had been supposed: 1,000 miles a year and as much as 50 miles in a single day.

Just before the end of the Grzimeks' pioneering expedition, Michael Grzimek was tragically killed when a huge griffon vulture collided with the right wing of their plane 600 feet above the Salei Plains, and the impact sent the plane crashing to the ground. The senior Grzimek went on to write a detailed account of their work, *Serengeti Shall Not Die,* and their published discoveries laid the groundwork for the ecological studies that continue today in the Serengeti. As a result of these studies, it is now known not only where the herds go but precisely what tissues of the grass they eat when they get there and why.

The rains generally begin in November in the Serengeti, and the zebras, wildebeests and Thomson's gazelles head for the open, shortgrass plains in the southeast. By January, hundreds of thousands of them have assembled in an almost unbroken mass stretching for miles. Other animals are nearby in lesser numbers: small herds of elands, the largest living antelope (the males weigh as much as 900 pounds); crowds of topis; and groups of long-horned Grant's gazelles, permanent residents of the plains that need almost no water to survive and remain after the others have moved on.

The ground is covered by a rich growth of shortgrasses with fine leaves that sprout from small cushions. Such grasses are well adapted to a land

A wedge-tailed eagle picks at the carcass of a kangaroo that was struck by an automobile in Australia's Northern Territory. Between sorties across the savanna looking for the rabbits and carrion that make up its main diet, this common eagle roosts in rockbound aeries and in treetops.

surface that is baked by day and chilled by night. The most common species here are dropseed grass, finger grass and star grass, and all remain short as a result of the vigorous grazing.

The animals, constantly on the move, seem to be indiscriminately cropping whatever is in their path; actually they are being selective, either consuming favorite species or eating separate tissues of the same grass species. The zebras, which differ from other grazers in possessing teeth in both jaws, bite off the taller, coarser part of the red oat and star grass and then move on. The lumbering wildebeests, with toothless upper palates, then arrive and chew the tender parts of the grass down to a point where it is quite short. After that, the 40-pound Thomson's gazelles take over; needing only 20 per cent of what a 350-pound wildebeest requires in one day, the Thomson's gazelles selectively nibble on the nutritious leaf tips at the base of the stems and feast on the small perennials growing between the tussocks. The grazing succession, as this pattern is known, also works in reverse. Wildebeests sometimes eat fresh grass from the time it sprouts

A harem of impalas cluster near a single male *(at far right, with horns)* on the open grasslands of East Africa. Seen from the flank, these fleet-footed antelope are a blended brown, but when they turn tail and run, black stripes on their hindquarters and black spots above their rear hoofs indicate the herd's direction, so that laggards can more easily follow the group.

until it is several inches high; later, when the grass is coarse and mature, the zebras come through to feed on the flowering sections.

After they are finished, the herds move to a different section of the plains, leaving the terrain looking like a sloppily mowed lawn. In two or three weeks they will return to take advantage of the new growth, maintaining this instinctive rotation until the wet season is over in May or June. Consequently, the grass is never overgrazed or trampled to the point where it cannot regenerate.

Outside the boundaries of the Serengeti preserve, the parts of the savanna where domestic herds graze offer a stark and sullen contrast to the luxuriant growth inside. The cattle, kept in tight groups by their masters within limited space, eat the ground bare and cut it to dust with their hoofs, killing roots of even the hardiest grass in the process. With the ground cover gone, the topsoil blows away on the gusty winds, and the land erodes, looking dry and lifeless even during the growing months.

The soil is also deprived of the cattle's manure; the herdsmen collect it in

stacks that they later burn or sell to farmers. Wild animals in the Serengeti, on the other hand, spread their manure over vast areas as they graze, returning the nutrients to the soil.

By June, the rains have finished, and the shortgrass plains have grown dry. The animal multitudes that covered the land have departed, making the trek of 100 miles to the tallgrass, acacia-woodland savanna in the moister western reaches of the Serengeti reserve, where they will spend the summer. The grasses there have grown tall, most of them to the height of an adult human, because there has been little heavy grazing by the animals who stay there year round. There is also plenty of fresh growth beneath the flowering stalks. The most abundant grasses of the western plains — and much of East Africa — are tall hood grass and red oat grass, mixed in with many other kinds: guinea, dropseed, bristle, shaggy-harp and the succulent-leaved signal grass. The plains are almost meadow-like, and the herds graze through them toward the wide valleys and rivers that lead to Lake Victoria.

When there is plenty of young growth, many wild animals will graze on the same grass. Recent studies have revealed, however, that given a choice of menu, the grazers demonstrate distinct preferences. In one area of the savanna it was observed that wildebeests went right for the stalks and leaves

Zebras and wildebeests mingle in a teeming mass on the Serengeti Plain, often feeding on different parts of the same plants. The constant movement of such herds enables even the most heavily grazed grasses to regenerate.

of red oat grass, while the Thomson's gazelles ignored the red oat and munched instead on star grass and guinea grass. Grant's gazelles, meanwhile, consumed grasses avoided by other grazers, and the topis chewed on dried stems and stalks even when fresh green growth was plentiful, thus clearing away the debris left by others. It is all amazingly efficient.

In the western parts of the Serengeti, where the great herds spend the dry season near the rivers that drain into Lake Victoria, numerous other herbivores are also present, all consuming the available food. Browsers, such as giraffes, elands and dik-diks, which feed primarily on shrubs and bushes, are part of the scene alongside the grazers. In the end, nothing is wasted, from the leaves atop the trees to inch-long blades of new grass and even underground bulbs, which the warthogs root out.

Each animal species contributes indispensably to the overall balance. The large herds of African buffalo, for example, requiring permanent shade and water, remain in the western and northern Serengeti all year. They feed on and trample large areas of tall, coarse grass, breaking down the harsh stems and mulching the ground. Thus when the migrant herds arrive, substantial sections of the plains are already prepared for them. Near the rivers, the heavy buffalo also eat and trample patches of grass that have been avoided by the equally huge hippopotamuses, which prefer to graze on the more tender grass nearby. Later, families of warthogs appear to dine on the grass eaten down by the hippos; the warthogs also use their tough snouts to break into thick tussocks in order to reach the nourishing and tasty bases of the stems.

Even the elegant, lofty giraffes, who usually ignore grass, make a vital contribution to the savanna's productivity. The tallest animals in the world (18 to 20 feet when fully grown), giraffes can feed at any height from ground level up. But they are surprisingly delicate eaters, selecting only food that has a high protein content. Though they occasionally bend down to pluck morsels from the ground, giraffes most often nibble on the green leaves of acacias as well as other trees and shrubs. They are especially fond of the ant-gall acacia, a tree that has bulbous growths teeming with ants at the base of its thorns; the ants and thorns repel most animals. The giraffes have tough, bristly lips, and tongues that can extend 12 to 18 inches. They pay no attention to the ants and eat new thorns, which are soft and supple; they avoid mature thorns, which are sharp enough to puncture automobile tires. Their feeding keeps the trees in check. In areas where large numbers of giraffes have been killed, the ant-gall acacias quickly become dominant, cutting into the productivity of the grasses.

The most prodigious consumer in the savanna is of course the elephant, which can weigh as much as four tons. Because it digests only 40 per cent of its food, an elephant spends most of every 24 hours searching for and eating the quarter ton of daily nourishment it requires. Elephants are major grass-eaters, devouring an average of 350 pounds per day, but they are also greedy consumers of other types of vegetation, feeding on as many as 134 different plant species. When times are hard, they gouge the bark from baobabs with their tusks to get at the supply of water inside the tree trunk. If the vegetation is out of reach, elephants simply push the trees over and eat what they want.

The elephants' mashing and husking of trees is nature's way of opening up new grazing areas, much as humans do with fire. As the giant pachy-

derms trample through dense thickets, light is let into dark woodland, and grass quickly follows. Elephants also help to expand the crucial water supply; their huge feet make indentations that trap rain water, and when they bathe, their vast bulk can turn large puddles into full-fledged water holes. During the dry season they dig for water in the beds of dry watercourses, creating a supply for others as well as themselves.

The dazzling array of mammals in the Serengeti — and on other African savannas — is complemented by the richness of the region's bird population. Birds play their own role in the ecosystem, feeding on insects, small rodents and reptiles as well as on seeds and grasses. Weaverbirds, which may breed in colonies of hundreds, depend on the grass not only for food but for material to use in building their intricately woven nests. If the rain is not sufficient to produce stems and blades long enough to weave and enough seeds and insects to feed their young, the weaverbirds do not breed at all — thus preventing overpopulation at a critical time. More than 200 other bird species feed mainly on insects; at least 60 kinds of hawks, eagles and vultures feed on live or dead animals by day; a dozen varieties of owls haunt the land by night.

Ground-dwelling birds also are common in the Serengeti. Among the most bizarre are the large black hornbills, with huge beaks, who walk sedately through the grass searching for insects, even digging up underground wasp nests. Tall secretary birds — so named because their crest feathers resemble quill pens — nest atop giraffe-pruned trees on large platforms made of sticks. Though secretary birds can fly, like the Australian bustard they prefer to walk, stalking for miles through the grass on stiltlike legs. They use their strong, clawed feet to stamp the ground to startle mice into moving and to deliver lethal blows to snakes of all sizes, including the venomous puff adder.

The most spectacular bird in the African savanna, however, is the ostrich. The largest bird in the world, the ostrich grows as tall as eight feet. It cannot fly, yet despite an average weight of 300 pounds, it can sprint across the grass at speeds up to 45 miles per hour. Ostriches are eclectic feeders,

African elephants feed among the white skeletons of acacia trees that they had destroyed earlier. Although the elephants subsist primarily on savanna grasses, they also are fond of the bark on trees and may wipe out an entire grove in their foraging.

A black rhinoceros chews a prickly branch that it has plucked from an acacia tree. Though the rhinoceros has keen senses of smell and hearing, it is severely nearsighted; this one depends on the vigilant red-billed oxpecker perched on its forehead to raise the alarm if a hunter or other danger approaches.

consuming not only grass and plants but anything small and shiny, including metal; one bird was found in South Africa that had swallowed 53 diamonds. Traveling in pairs or in flocks of 10 to 50, the ostriches often intermingle with the great herds of zebras and wildebeests, where their acute hearing and extraordinarily sharp vision make them valuable sentinels. Their ability to detect danger at a great distance serves as an early warning to the grazing herds that predators are approaching.

A pair of ostriches, the male with glossy black plumage and the female with grayish brown, stalk the grassland searching for seeds and insects. Indigenous to the Serengeti Plain, the flightless ostrich sometimes hides from predators by lying on the ground with its neck outstretched — behavior that gave rise to the myth that the ostrich buries its head in the sand when frightened.

Wherever there are herbivores in the savanna, there are flesh-eating predators — lions, leopards, cheetahs, hyenas, jackals, foxes and wild dogs — searching for their own food supply. The predators' attacks serve to keep down the population of the herds, which might otherwise increase beyond control and overgraze the savanna. The lions, the biggest and most powerful of the predators, kill the biggest prey: large antelope, zebras and even young rhinos. Lions also have a special predilection for the meat of warthogs and will wait for hours for them to emerge from their underground nests. Warthog mothers, in turn, are so fiercely protective of their families that they are known to turn on their tormentors, whatever the size. A game warden witnessed one warthog attack a leopard and another rush an elephant; the larger animals were so surprised that they fled in panic.

Leopards, who hunt by night, and cheetahs — fastest of all animals over short distances — mostly go after gazelles and impalas, the reddish brown antelope whose prodigious ability to leap 30 feet in a single bound is an effective defense. Hyenas, which sometimes prey on the newborn calves of wildebeests and even on old and sick lions, are also scavengers. Their immensely strong jaws enable them to crush the largest bones of animals killed by others. Jackals and vultures come in to feed with the hyenas; ants and other insects then complete the job, leaving the plains clean of carrion. Once again, nothing is wasted.

The great herds generally linger in the well-populated western valleys of the Serengeti until August or September, while scorching heat continues to bake the eastern plains. As the land becomes drier, the herds head north to the Mara River and into Kenya. By October, the whole savanna is at its hottest and driest, the fires are burning and anxiety is mounting again about when — and if — the rains will come. The female wildebeests, which have been carrying their unborn calves during the dry months, feel a special urgency to find fresh grazing in order to produce the milk that their young will soon require.

At length the first storm front approaches, with storks and other harbinger birds gliding high on its updrafts. The animals of the Serengeti see the storm at a distance and move toward it, grunting and braying, their hoofs pounding the hard ground in a thunderous rumble. The rains have returned, and the life cycle of the savanna begins anew. Ω

A crested crane, named for the golden halo of bristly feathers on its head, inspects the clutch of eggs in the straw nest it has built on the savanna. The male and female cranes take turns at the 30-day job of incubating the eggs.

A fierce lappet-faced vulture swoops in to claim the first portion of an animal carcass, while smaller Ruppell's griffon vultures wait their turn. With its power

ak, the lappet-faced vulture will slash through the dead animal's hide. Only after this cut is made can weaker scavengers get at the flesh within.

Chapter 4

TUNDRA: THE HARDY, DELICATE LAND

Canada's Yukon is renowned as a treasure house of precious metals. But gold and silver are not the only objects of great value to emerge from the frozen ground of that wild land. Whole bodies and skeletons of prehistoric animals have been unearthed there, disinterred during placer mining operations by the powerful hydraulic hoses used to flush away the dirt around gold-bearing ore. Among the earth-stained remains that have thus come bubbling to light are the bones and carcasses of woolly mammoths, bison, wild horses and other assorted beasts that once roamed the Far North.

One of the most amazing discoveries seemed at first little more than a curiosity: a series of tiny, interconnected burrows containing the fragile bones of the collared lemmings that inhabited them and the rodents' winter larders of seeds. Some sudden catastrophe had overtaken the animals; perhaps the tunnels had been buried by a landslide. The effect, in any case, was to entomb the lemmings and their caches 10 to 20 feet below the surface, and then, as the icy ground froze permanently, to preserve them.

The mining engineer who spotted the burrows before they were destroyed took one skull and a handful of the larger seeds home as souvenirs. Fortunately, he kept the seeds in a dry spot, where they would not deteriorate, for 12 years would pass before his little collection received the scrutiny of scientists from the National Museum of Canada. The seeds proved to be those of Arctic lupine, a perennial herb that erects spikes of bright blue flowers early each summer after its first three years of patient growth. This information was in itself of no particular import, but the subsequent dating of skull and seeds was. They were at least 10,000 years old, and some of the seeds looked as hard and shiny as the day the lemmings gathered them. Apparently, the animals and lupines had flourished throughout the last great ice age in one of those unglaciated areas known as refugia.

Impressed by the seeds' freshness, the scientists decided to experiment. They placed the best-preserved on wet filter paper in laboratory dishes. Within 48 hours, six seeds had sprouted. Planted, the seedlings kept growing, and in less than a year one had bloomed, the delicate flowers a startling reminder of an Arctic summer that had come and gone when most of the Northland was covered with thick ice. Ancient seeds have been found elsewhere, but nowhere else have seeds this old germinated. Adapted to survive in the cold, they exemplify the hardiness of tundra species.

A kind of prairie, the treeless tundra covers a tenth of the earth's surface, in an irregular band around the top of the world. Beginning where forests end and sweeping to the polar seas, it stretches from Greenland through

The Arctic lupine, a hardy perennial herb, blossoms in the tundra during northwest Canada's brief summer. A member of the pea family, the lupine provides food for grazing animals and helps other plants to grow by adding nitrogen to the tundra's acidic soil.

A small ground upheaval in northern Alaska reveals an ice lens formation. The ice lens was created underneath the surface when water ran into a crack in the soil, spread out on top of the permafrost and then froze.

Canada, across Alaska and Siberia, into Scandinavia. The tundra is not a true grassland since grass covers only part of it; in places it is barren. It is a harsh realm of limited food resources, high wind and bitter cold for much of the year, subject in midwinter to continuous darkness as muffling as the snow that blankets the land. But surprisingly in this region so close to the North Pole, the snows are not usually deep. In fact, annual precipitation is only 12 to 20 inches. Yet there is no lack of water.

The perpetually frozen ground, or permafrost, beneath most tundra acts as a shield, impeding drainage. Almost 85 per cent of Alaska and 50 per cent of Canada and the Soviet Union have this frozen underpinning, and in some areas it is almost a mile deep. Since the top layer of soil rarely thaws below a few inches, runoff has nowhere to go; it can only form countless lakes and ponds and large bogs, where grasses and sedges often grow.

Even with the approach of summer and 24 hours of daylight, the soil temperature seldom rises much above the freezing point; this is because so much of the sun's radiation goes into the melting of ice and snow and the thawing of the top layers of the permafrost. Tundra plants must not only be capable of carrying on their life functions under cold conditions; they must do so in the short space of the Arctic growing season, two to three and a half months long. In this demanding environment, plants and animals have become masters of survival, and their adaptations are various and fascinating. But when compared with their grassland counterparts in warmer climates, the numbers of tundra plant and animal species are few.

Unveiled when glaciers withdrew from much of the land 8,000 to 15,000 years ago, this youngest of the world's regions is still being shaped by ice today. The tundra may stretch endlessly as plain or be rolling or mountainous, but everywhere it includes many of the same unique, ice-produced features. Among its prominent and characteristic landforms are pingos, great mounds of ice covered with earth and plants, and polygons, whose ice-filled edges form honeycombs that pattern the ground for miles on end.

Pingos can be seen rising unexpectedly from the flatness of the plains, miniature hills often ruptured at the top to reveal their icy cores. They may be as much as 150 feet high and 1,800 feet across at the base. Formed when water is locked in by permafrost and freezes, they thrust themselves up from the tundra in the same way frozen milk lifts the cap off a bottle.

An unusually large pingo, an ice mound covered with earth and vegetation, rises 150 feet above the flat surface of Canada's MacKenzie River delta. The pingo emerged when an ancient lake dried up; the ground beneath the lake bed, having lost the insulation of the lake's water, froze and expanded prodigiously.

The irregularly shaped polygons are the result of a different process. They may be 10 to 100 feet wide. They develop when the soil contracts in response to changing temperatures, with cracks opening like those in the bed of a dried-up pond. During summer the cracks fill with meltwater, which later freezes into wedges of ice. The process repeats itself from year to year, and the wedges — which underlie the perimeters of the polygons — gradually grow. With the spring thaw, the cracks become water-filled channels at the junctures of which small ponds often form. Such a system of interlocking channels has been well described by scientists as a beaded stream. Indeed, from the air the pools can look like glittering strings of beads.

Often these pools become large ponds as their banks thaw and slump into the water. Arctic lakes also grow like this; sometimes two will meet and coalesce into a larger body. But while the lakes may expand, they do not usually deepen, since dissolved material from their banks continually fills their beds. Many are only 5 to 10 feet deep, and the deepest may be no more than 20 feet. The largest tend to be elliptical and to be oriented along their axes at right angles to the prevailing wind, which apparently drives the current and accounts for their shape. Over time, many of the lakes and ponds disappear. Some drain: Erosion cuts openings that let the water escape. Others fill gradually with vegetation: Pendant grass (a highly nutritious plant related to Kentucky bluegrass) grows out into the water, rooting in soil from the collapsing banks, and eventually takes over.

Interlocking polygons of ice etch a wintry pattern on Alaska's North Slope. During the summer, cracks in the soil fill with water that later freezes into a honeycomb of vertical ice wedges. The wedges grow marginally larger each year in a process that may continue for several thousand years.

The constant cycle of freeze and thaw contributes to yet another feature of the tundra, its unevenness. Small mounds, or hummocks, can make hiking here an ankle-wrenching experience. They are the result of thawed, poorly drained soil being squeezed between a freezing top layer and the rock-hard permafrost below; with nowhere to drain as the pressure builds up, the soil erupts as a blister. In this manner great portions of tundra are turned over, and the soil is mixed as effectively as by any plow.

A similar phenomenon — frost heaving — forces rocks to the surface and sorts them so neatly that it seems a human intelligence has been at work. For reasons still not fully understood, the rocks move out into circles, with the bigger stones shifting to the edge. Many such contiguous circles form a kind of net, but on a hill or mountainside, the stones may be sorted in parallel rows, strung out downslope by the force of gravity.

Gravity is responsible for another phenomenon of the tundra. Water-

Alpine Tundra in the Temperate Zone

Above the tree line and below the cap of perpetual snow on almost all the earth's high mountains lie bands of tundra created by altitude rather than latitude. These zones of alpine tundra cover a total of four million square miles, most of it in the North Temperate Zone.

Unlike tundra at high latitudes, which is deprived of sunlight by the long Arctic winter, alpine tundra receives daily doses of solar radiation. And while little snow falls in much of the Arctic, alpine areas are subjected to some of the heaviest snowfalls on earth.

Despite these climatic differences, the vegetation in alpine tundra has much in common with its cousins in the Arctic. It consists mostly of stunted plants, often widely separated by bare soil or rock. The species that survive are those hardened to short growing seasons and chilly summers.

Alpine tundra also is home to strong-winged birds and such sure-footed mammals as the Rocky Mountain goat, the Andes llama and the Swiss Alps ibex. When winter storms become too much for even these heavy-coated species, they can simply migrate down to the tree line to find food and shelter.

logged soil covering an incline may surge forward, sliding over the permafrost to produce a series of lobes, terraces or stripes. This process, called solifluction, can entrap animals or bury them; it accounts for the discovery of many complete animal remains in the Arctic.

While ice and cold have a bearing on the look of the land, they also directly affect the growth patterns of tundra plants, even the familiar grasses and sedges. Most tundra vegetation is so stunted that it seems to cower. Low-growing willows cling to the tundra, poking their branches 10 to 15 feet across the soil but never raising them more than a few halting inches. Dwarf birches may stand only a foot high, yet analysis of their growth rings may show them to be 100 or more years old. Such retarded growth is largely the result of permafrost, which prevents the plants from developing deep root systems and locks up the soil's nutrients in ice. But poor drainage and inadequate aeration of the acidic soil are also factors.

More than 99 per cent of tundra vegetation consists of perennials— grasses and sedges, small flowering plants, mosses and lichens and shrubs like the ground willow. Annuals simply do not stand much of a chance; the growing season is too brief for them to sprout, flower and go to seed before autumn arrives. One way the perennials compensate for summer's brevity is to keep most of their living tissue safely belowground; through the winter they store in roots, rhizomes, corms and bulbs the energy and nutrients they will need to initiate growth in spring and to reproduce. As soon as the melting snow signals the start of the short growing season, they tap these reserves and expend them in a few frantic weeks on flowers. Some plants are so opportunistic that they form partial flower buds and leaves almost a year in advance to be ready for spring's first stirrings. The evergreens need not immediately channel their energy to the development of new leaves but rather can use it on blooming. The grasses and sedges themselves have a semi-evergreen cycle; although they may appear to die back to the ground in the autumn, as do their relatives in more southerly regions, they preserve at their cores the green shoots that will become next year's foliage.

During flowering, the plants consume their energy reserves faster than they can replenish them. But this expenditure is made up for later, when additional foliage can be produced to aid in photosynthesis. In keeping with the economy that characterizes tundra life, the ripe seeds tend to be small, without burred or fleshy fruits, ready for distribution by wind or animals. And some are so specifically adapted that they will germinate only in a warm spring, the plants' way of ensuring that the seedlings will stand a fair chance of actually growing to maturity.

Although they reproduce sexually, depending on wind or insects to pollinate them, the plants would be ill-served if this were their only means of perpetuating themselves. In a particularly cold, bleak summer they may fail to set seeds at all, and many years may pass without successful sprouting. Thus it is important that the plants also be able to reproduce asexually. The grasses and sedges accomplish this with the aid of their rhizomes, those underground stems that spread and send up shoots of their own, or through layering, in which a branch or a stem touches the earth and takes root. Other plants grow bulbils on their stems; these tiny bulbs are dispersed by passing animals or drop to the soil where, with luck, they germinate.

The new plants can, under favorable summertime conditions, draw upon

Patches of yellow locoweed, white phlox and purple lupine blossom in July on a stony slope 500 feet above the tree line in the Olympic Mountains of Washington State. In this zone of alpine tundra, midsummer temperatures reach 55° F. by day but drop near freezing at night.

the dead tissue of the parent plant for a boost. Low temperatures keep the old leaves and branches from rotting easily and create a build-up of raw humus or peat, which serves as a kind of nutrient bank for the living vegetation. When attacked by fungi, the dead matter breaks down, and its riches are at the disposal of the plants that deposited it in the first place.

To take advantage of the summer sunlight (available 24 hours a day for three and a half months at lat. 75°), many species have steeply inclined foliage that catches the rays as they stream in at a low angle. In another adaptation, several plants have leaves that tend to be large for the plants' size so that they may capture as much of the radiation as possible.

One of the more impressive features of tundra vegetation is the ability of many plants to create their own microclimate. Arctic cotton grass (so named because it produces a pompon of purest white fluff) grows as a tussock that looks like an uncombed wig. Within such a shaggy mound, temperatures can be higher than in the air outside. As more shoots develop, they push the older ones outward. The tussock's thick growth insulates the earth beneath it, keeping the soil from immediately freezing with the onset of autumn. Then, as soil around the plant's base congeals and expands, it crowds the still unfrozen portion, causing the ground to well upward and lift the tussock. When spring comes, the tussock not only stands taller but can receive more of the sun's rays. Thanks to the warming it undergoes, the tussock emerges from the melting snow 4 to 10 days before its lower-lying neighbors. And this happens at the most advantageous time, the period of the summer solstice, when the sun's angle is highest. The plant's growing season is thus 5 to 10 per cent longer than that of adjacent species.

The crimson leaves of the Arctic bearberry plant add splashes of color to a lichen patch during the frosty last days of summer. A ground-hugging shrub, the bearberry produces berries that are eaten by birds as well as bears.

A survey of the tundra reveals other temperature adaptations. Some plants show hollow stems, which function as miniature greenhouses; they may be 36° F. warmer within than without. Hollow stems are also economical; they save the plant from producing a core, thus conserving valuable resources for more critical purposes. Other plants wear a kind of fur coat; the woolly lousewort encloses its stems and buds in a mass of fine fibers. Still others grow in rosettes or in cushions that not only heat up in the sun but deflect the wind as well. (In such cushions, the wind velocity may be reduced by as much as 99 per cent.) Where many different plants grow together in a mat, they form a low-lying canopy whose temperature may be several degrees higher than that of the air just above it. The parabolic shape of some tundra flowers, such as the buttercups, focuses heat on the developing reproductive parts; the flowers turn their faces slowly, following the sun on its journey across the sky. On cold days, insects often shelter within the petals and, so protected, unwittingly pollinate the blossoms.

Even the color of a plant may enhance its temperature. The darker the leaves, the more they are able to absorb the sun's rays. Several species produce a red pigment that combined with their green chlorophyll makes them appear almost black to the eye. Growing in a compact community, they may be as much as 28° F. warmer than the surrounding air. Others with dark coloration utilize the filtered light reaching them through snow to photosynthesize, thereby creating heat that becomes trapped around them. In this snug environment, they may get a two-week head start on the growing season, a vital edge where summer is so short.

Tundra plants must guard against the wind. During winter, when the moisture they need is frozen in the soil, they stand in danger of desiccation

PURPLE MOUNTAIN SAXIFRAGE (2 X LIFE SIZE)

Purple mountain saxifrage grows in tufts, or cushions, only an inch or two high in moist, rocky regions. The close stacking of its many small leaves helps increase the plant's ability to photosynthesize efficiently during the tundra's brief summer season.

Each summer, myriad wild flowers brighten the tundra's drab complexion with a blush of color. As winter cold grudgingly surrenders to the warmth brought by longer hours of sunlight, ground-hugging perennials produce flowers whose beauty is intensified by the plants' diminutive size.

The speed at which the flowers begin to appear is itself remarkable. Well-acclimated to the tundra's brief growing season — often no longer than 50 days — the plants burst forth with new growth at the first signs of spring. The purple mountain saxifrage shown above, for example, can grow, flower and produce seeds within a month.

The flowers themselves are well suited to the tundra's uncertain weather. Despite their fragile appearance, they are amazingly resilient. A researcher working in northern Canada recorded a spring thaw during which a willow began to flower. The temperature dropped dramatically, freezing the willow's buds solid for three weeks. Once true spring came, the catkins resumed the process of opening and producing seeds as though nothing had happened.

Low temperatures limit to a few mild weeks the activity of the relatively few flies and other insects that pollinate most tundra flowers. However, some scientists believe that the flowers' showiness, by catching the flies' attention, may increase their likelihood of being pollinated.

Artist Andie Thrams of Anchorage, Alaska, has captured the delicate charm of seven of these wild flowers in the drawings shown here and on the following pages. Each summer, she journeys across the Alaskan tundra, identifying plants and sketching them from life.

WOOLLY LOUSEWORT (.6 X LIFE SIZE)

MOUNTAIN AVENS (LIFE SIZE)

ARCTIC POPPY (.5 X LIFE SIZE)

ARCTIC WILLOW (LIFE SIZE)

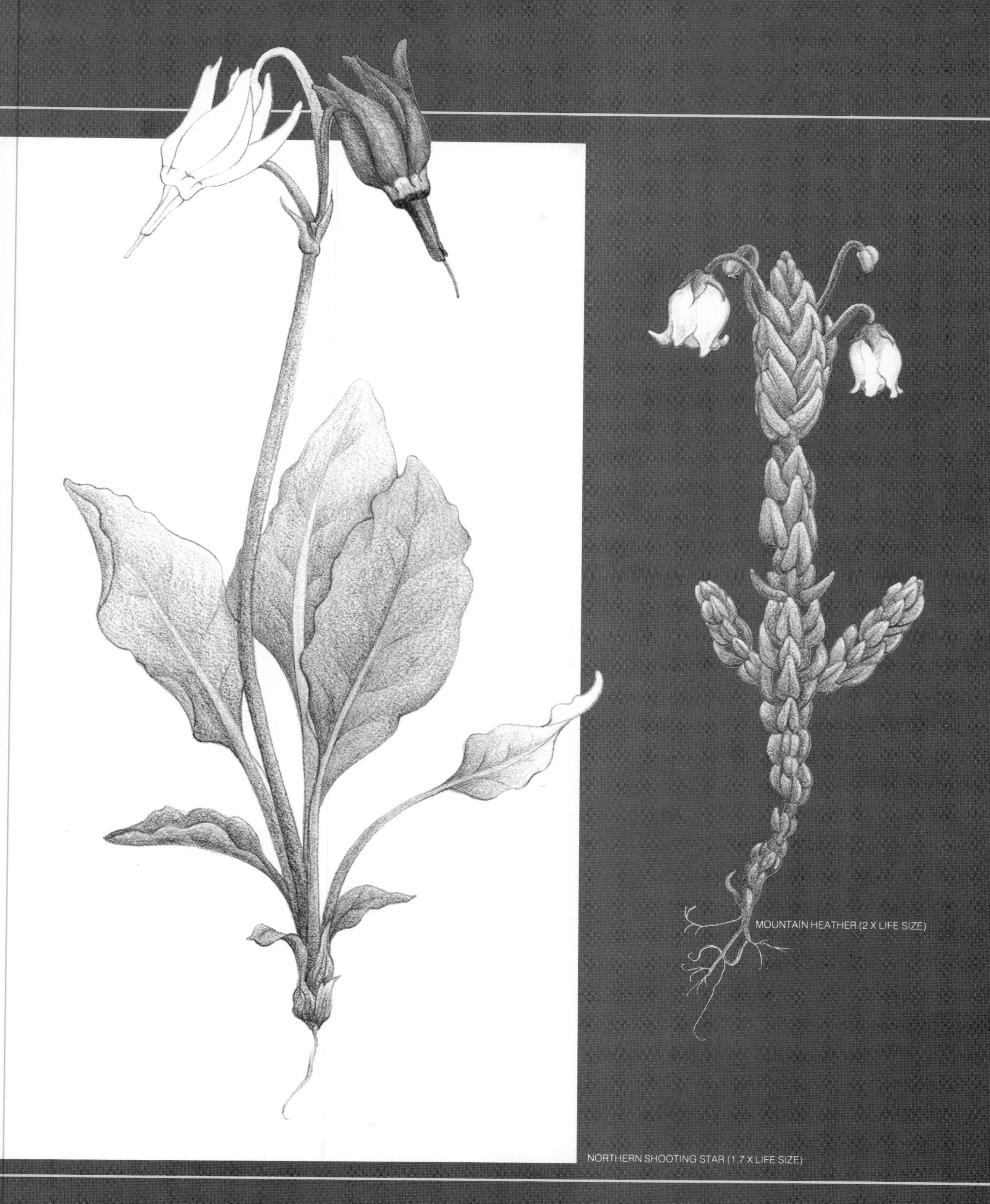

MOUNTAIN HEATHER (2 X LIFE SIZE)

NORTHERN SHOOTING STAR (1.7 X LIFE SIZE)

and may be severely abraded by the sharp-edged ice crystals flying about. Plants that retain their leaves are necessarily thick-skinned. But even in summer the wind can dry out plants growing in marginal areas, and many have leathery or waxy surfaces. Some replenish moisture by soaking up meltwater through their leaves. And where wind and cold have killed portions of a plant, the dead branches can act as a buffer, breaking the force of the wind and causing eddies that tuck snow protectively around the plant.

The tundra's most delicate-looking plants are often the hardiest. The Arctic poppy raises its flower — bright yellow, instead of red — on a spindly stem that whips about in the wind, yet the tissue-thin petals do not fall until ready. The velvety mosses, soft and cool to the touch, can withstand drought conditions, withering without dying. And the seemingly frail lichens — so brittle on dry days that they crunch underfoot — can reconstitute themselves quickly after rain begins. One variety, called the staghorn because of its antler-like outgrowths, rolls up in a ball as it dries and is blown across the tundra until it absorbs enough moisture to unfurl again.

Lichens are primitive plants, without stems, leaves or roots. Actually, they are two plants rather than one — an alga and a fungus living together in an arrangement beneficial to both. The alga, which contains chlorophyll, manufactures food for the fungus; the fungus in turn brings the alga water and mineral nutrients, soaking them up through cell walls thick enough to protect the alga from excessive sunlight. Lichens grow on rocks, wood and the ground, often in the most marginal areas, where other plants could not survive. Those on rocks perform a basic function; they secrete acids that dissolve some of the stone and contribute thereby to soil building.

Lichens can be very old (some found in Greenland have been alive for more than 4,000 years), and they expand slowly, many by only one fifth of an inch a year. However modest their growth, they can spread over wide patches of soil to form thick white, creamy or gray mats. Stepped on, they may shatter, but pieces can become new plants, providing the fragments contain both fungal and algal cells.

The parabolic shape of a buttercup's petals *(top)* focuses the rays of the Arctic sun so that the plant's developing seeds are several degrees warmer than the surrounding air. This pocket of warmth also attracts the flies and other insects that pollinate the buttercup. The spiderplant *(bottom)* reproduces by sending out slender runners that find hospitable niches in the barren ground, take root and grow. The new plants receive nourishment from the mother plant until they are able to exist on their own; then the runners disintegrate.

In a realm as exposed as the tundra, animal life is restricted. Yet despite the limitations, some species flourish, especially where vegetation is thick or forms a ground cover that will bear browsing and grazing. And while the populations may not be particularly diverse, some of the animals occupy niches similar to those filled by their grasslands counterparts.

Arctic ground squirrels, or parka squirrels, thus named because Eskimos use their skins for parkas, look much like prairie dogs and dwell in similar colonies, surviving on a diet of leaves, roots, berries and seeds. But they can live only where the soil is well drained, as in hills, riverbanks and accumulations of sand; here the absence of permafrost permits the squirrels to excavate burrows for their winter hibernation. Moreover, they dig where the snow will pile deepest; this insulation will prevent subterranean temperature from dropping below 10° F. while they sleep away the winter. Investigators wondered how ground squirrels could withstand such cold until they examined the burrows and discovered that the entrance to the nesting chamber always lies below it. The warmth emanating from the squirrels' bodies is trapped in the chamber. And since the surrounding soil is well drained, there is little or no ice for the heat to melt, with the result that the chamber stays dry. Hibernating an incredible nine months, the rodents

Flowering plants have adapted to the days of continuous sunlight north of the Arctic Circle and to the cruelly cold, sunless days that follow. The stem and buds of a Jacob's ladder *(top)* are enclosed by fibrous hairs that protect the plant from overexposure both to the sun and to wind and cold. The Arctic forget-me-not *(bottom)* grows in a cushion-like cluster whose streamlined shape greatly reduces the desiccating force of winter winds.

emerge from their dens in early May. In the frantic days that lie ahead, they must somehow breed, raise their litters of 5 to 10 young and build up enough body fat to sustain themselves through the long winter.

Even more amazing are the most numerous of the tundra mammals, the supposedly suicidal lemmings. In Scandinavia, these mouselike creatures, no more than five inches from tip of nose to end of stubby tail, are periodically seen advancing in hordes. A few years ago, so many descended on a town in northern Norway that the dogs soon tired of killing them. Thousands swarmed over railroad tracks just as a locomotive approached, and the ensuing slaughter left the wheels so bloody that they could not maintain traction. In a later invasion, the animals' advance was hindered by a fjord. But the water did not stop the single-minded lemmings from attempting a crossing. Some began to swim, only to drown. Others scurried over a bridge, but it became so crowded that those in the middle panicked, scampered to the edge and jumped off. Those that struck the ground died. Those that plunged into the water were swept away by rapids.

Episodes like these helped give rise to the old myth that lemmings commit suicide. In fact, the frantic lemmings were engaged in a lifesaving activity — a mass migration to escape crowding. Interestingly, only Scandinavian lemmings migrate in such numbers. But all lemmings can be subject to the stresses of overpopulation. For various reasons, lemming numbers peak and then crash, and years when the rodents occur in great abundance on the tundra are followed by years when they are scarcely to be seen. This boom-and-bust cycle led some early Scandinavians to believe that lemmings fell from the sky with rain and that, after consuming all green things, they were devoured in turn by larger beasts. The Alaskan Eskimos also thought that the rodents came from on high; their word for the collared lemming translates as "the one that fell from the sky."

Lemmings are breeding machines, and they go on mating throughout the greater part of the year, the only animals in the Arctic to do so. Instead of hibernating like the ground squirrels, they spend the winter under the snow, feeding on the green shoots of the sedges and grasses. They use tunnels and runways to get around and sleep in grass nests. Occasionally, they come to the surface through snow chimneys and scamper about.

A healthy female surviving to a ripe old age of two years could produce as many as 14 litters of seven to eight pups each. Few lemmings live that long, however; most fall victim to predators and to the spring thaws that flood their runways and burrows. But even over a relatively short life span, a female can still bear several litters. It takes lemmings only four days to double their birth weight, and young females may become pregnant even before weaning. Over the course of a single winter, providing conditions are favorable, a lemming community can increase its population a hundredfold.

To sustain themselves, lemmings must consume up to twice their weight in food every day, and their consumption of grasses and sedges can have a direct effect on the level of permafrost and on the quality of the vegetation. They eat quickly, foraging 14 times a day, and are inefficient digesters of their food, with about 70 per cent of what they take in deposited on the ground as feces and urine. When spring comes and the snow melts, the wastes soak into the soil, fertilizing it and thus contributing to the vigor of the plants. The lemmings enrich the environment in another way. During winter, they clip dead vegetation to get at the green shoots hidden at the

Two varieties of lichen grow side by side on a barren rock surface in northern Canada. The yellow-orange crustose lichen clings to the rock in a tight crust, while the dark, leaflike plants, known as foliose, are each attached to the surface by a single umbilicus. Both of these varieties lie dormant through long dry spells and then quickly absorb rain, dew or any other kind of moisture when it becomes available.

A third major kind of lichen, the shrublike fruticose lichen *(near right)*, sprouts tiny branches whose erect stance and red caps inspire their popular name, British soldiers.

The thin strands of Alaskan haircap moss, barely two inches tall, tower above a tiny bog cranberry plant *(far right)*. Mosses, like lichen, have no roots; they are anchored to the tundra soil by rhizoids and require only minimal warmth and moisture to survive and proliferate.

base of the tussock, thus hastening the process by which the lifeless stems and leaves decay and return nutrients to the earth. Areas where lemmings have foraged can rank among the lushest on the tundra. But their heavy cropping can remove so much of the insulating ground cover that the sun's warmth is able to penetrate deep enough to thaw some of the permafrost. Such denuded areas may take several years to recover.

There is a hazard in all this for the lemmings. As their numbers increase in response to an abundant food supply, predatory animals move into the area. A year of abundant lemmings is a year of feasting and high reproduction for Arctic foxes, weasels, snowy owls and gull-like jaegers. Wolves, too, may divert their attention from larger prey to feed on the lemmings. Even living under the snow in winter does not afford the lemmings complete protection. Arctic foxes are able to detect lemmings through the snow, and they dig them out. When the lemmings come up through their snow chimneys for air, they are often picked off by a snowy owl. And weasels, whose long, slim bodies enable them to slither through the tunnels in search of a meal, often trap the lemmings at home. Lemmings are most vulnerable in spring. Not only can they drown in their burrows during the two-week runoff period; they risk getting their coats wet. When this happens, their fur loses its insulating properties, and many die of exposure. Worse, with the snow melted and the vegetative cover reduced by grazing during the preceding winter, they are easy prey for predators.

When the lemming population crashes — as it does every three to six years — the predators feel the effects. Those dependent on the lemmings for food either die off too or migrate. No one knows precisely why these drastic declines in lemming numbers occur. Obviously, the weather and predation play a large part, but other factors seem to be involved as well.

The rigors of many lemmings living in close conjunction and competing with one another for food may lead to adverse endocrinological reactions that inhibit breeding. A reduction in the quantity and quality of the food supply may also contribute to stress. It has even been suggested that as the lemmings remove more and more of the ground cover and the depth of thaw

increases, roots penetrate to levels where the nutrients are so diluted that the vegetation suffers. Obliged to consume this less nutritious food, the lemmings suffer, and their chances of breeding successfully decrease. Whatever causes the cycle, the boom-and-bust process is repeated endlessly on the tundra. And in the interim, while the lemming numbers are down, the vegetation recovers, and the permafrost level once again rises.

The absence of lemmings does not spell a summer of desolation for the tundra. Insects, which find the ponds and bogs ideal for breeding and provide food for the fish that inhabit the larger bodies of water, are out in the billions, and birds are present in the millions. Geese, swans, ducks, gulls, waders and songbirds can be seen nesting everywhere, suggesting in their numbers what the avian population must have been like on the American prairies and plains before plows turned the sod and waterways were opened to commerce. And since there are no trees for them to nest in, most do so on the ground, where they are fair game for predatory mammals and such rapacious birds as owls and jaegers. The oldsquaws, one of the 20 species of duck that migrate to the North in spring, have taken to nesting among Arctic terns, which are fiercely protective of the area around their own nests and through their raucous, defensive behavior alert the ducks to danger.

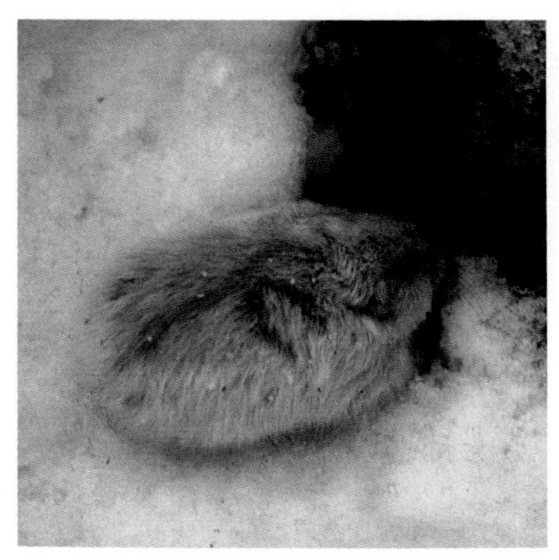

The winter coat of a collared lemming blends with the snow-covered Arctic tundra, protecting the lemming from larger mammals and birds that prey on it. The populous lemmings spend the winter tunneling beneath the snow in quest of willow shoots and sedges.

Many of the migrants feed on the abundant insects, while others concentrate on fish, aquatic vegetation or tundra plants. The fish-eating loon is an excellent diver, but because of its set-back legs, the bird has a hard time moving about on land. It is a strong flier, however, and has a habit of taking to the air at dawn and sweeping in circles above the water. "This chore of combined morning exercises and countryside rousing completed," write the authors of *The Birds of Alaska,* "the great diver glides back to the water and starts getting breakfast, stopping now and then to give way to great peals of laughter at the excellence of the food, the joy of being alive, or perchance at having awakened all the furred and feathered folks, with his greeting to the rising sun." The red and the red-necked phalarope, specialized shore-bird species, feed on aquatic insects and plankton in lakes and ponds. They resemble mechanical birds as they spin on the water, creating little whirlpools that bring food to the surface.

Grass-eating geese are so numerous in some areas that they can actually affect the appearance of the land through their food habits. This was observed by Soviet scientists studying bird life on the Taimyr Peninsula of northern Siberia. Here geese consume whole patches of cotton grass, eating everything — the rhizomes and starchy nodules on the roots, as well as the leaves and shoots. When they have finished, little else than moss is left. The removal of the cotton grass lets more light reach the soil, thus warming it and thawing the permafrost. The geese's excrement adds nitrogen to the soil. It is no accident that where geese and other birds congregate on the tundra to nest, feed or molt, the surviving or colonizing vegetation grows vigorously. Canadian scientists have discovered that in addition to fertilizing the soil this way, marsh geese pluck enough grass to encourage the growth of blue-green algae on the ground. The algae are nitrogen-fixing and further enrich the soil with the production of this essential element.

The birds begin arriving in the Arctic in early spring. First on the scene are the waterfowl, drawn to open patches in the ice. They are followed by the shore birds and finally by the land birds. Some of the visitors come from thousands of miles away. The Arctic terns, for example, journey from as far

Only its dark eyes and nose reveal the presence of a short-tailed weasel, or ermine, among the snowdrifts of Lapland. In summer the weasel's pelt is brown, but as winter approaches it molts into its coat of snow white.

south as Antarctica and make a round trip of 21,000 miles every year. All are driven by the same imperative: to mate and nest before the short season has advanced too far. And they accomplish this against great odds; foul weather may reduce the amount of food available. On a hurried timetable, the brant goose often mates before arrival. The sanderling produces two clutches of eggs at a time, as a kind of insurance against loss. The female sits on one clutch, the male on the other, and when the eggs hatch, each bird takes care of its own brood. Such behavior is in distinct contrast to that of the red-necked phalarope; after the female lays her eggs, she abandons the nest and the male takes over, fulfilling all parental duties. Interestingly, the emancipated female — not the male — is the more brightly colored.

Of the 100 or so species that nest on the tundra, only a handful spend the entire year there. Among these hardy denizens are the raven, the snowy owl and the rock ptarmigan. The raven winters over by eating the remains of dead animals, droppings and garbage from human settlement. The gyrfalcon, the world's largest falcon, catches Arctic hares and ptarmigan. And the snowy owl, swooping down on silent wings, takes a variety of prey, including, of course, lemmings. It is so distinctly a northern bird that it nests and lays eggs while there is still snow on the ground.

Ptarmigan are by far the most numerous of the winter holdovers. Unable to scratch through the snow to locate their food, they often cluster where the snow cover is thin and feed on exposed berries, seeds and the dormant buds or twigs of dwarf willows and birches. Their plumage turns almost completely white with the onset of autumn, and thus they are well camouflaged in winter. The loss of pigment that occurs with this color change leaves empty spaces within their feathers; these trap body heat and help insulate the birds. During particularly severe weather, the birds burrow into snowbanks to keep warm. In another winter adaptation, ptarmigan grow a cushion of feathers on their feet that keeps them from sinking into soft snow. When spring comes, the female turns speckled brown and, thus rendered inconspicuous, sits safely on her nest. The male, which actively assists in the rearing of the chicks, is slow to revert to summer plumage.

Birds are not the only summer visitors to the tundra. In the spring, the North American caribou and the reindeer of Europe and Asia migrate there from the forests where many shelter during winter. They are, in ecological terms, the tundra equivalent of the pronghorn of the American prairies or the Thomson's gazelle and the wildebeest of Africa's grasslands; but unlike those creatures, they have little competition for food. Both caribou and reindeer are admirably suited to life in the North and are, in fact, the same species, *Rangifer tarandus.* They have big, concave hoofs that keep them from sinking deeply into the snow or marshy tundra; a thick coat with hollow hairs, whose air spaces provide insulation by capturing body heat; and a mouth that can sort living plant tissue from dead and discard the indigestible bits. Bulls weigh an average of 350 pounds and females 200.

Although members of the deer family, the caribou and reindeer differ from most other deer in that both sexes have antlers. The males' are larger and can measure more than four feet across. But whereas the bulls shed their antlers soon after the rutting season, the cows do not do so at least until winter; pregnant females wear antlers until after the spring calving. With their antlers, the pregnant cows can drive other caribou or reindeer from foraging grounds that they have scraped free of snow.

Caribou and reindeer are almost always in motion, traveling in large herds during autumn and spring. A herd on the move is one of nature's most splendid sights. For three days Canadian scientist John P. Kelsall watched spellbound as 30,000 caribou marched by his camp (actually not a large number, for herds can be hundreds of thousands strong). "The feed-

A red-throated loon peers from its nesting place of sedge next to an Alaskan pond. The migratory loons live in isolated pairs and fly south when their ponds begin to freeze.

A snowy owl shelters its young in their shallow nest on the ground. A skilled hunter of lemmings and other small rodents, the snowy owl is one of the few birds that are able to survive year round in the Arctic tundra.

ing animals covered the surrounding hills," he noted with awe. "Sometimes they literally stumbled over our tent ropes. We lay motionless on the open ground and found that even downwind of us caribou would pass within 10 to 15 feet. Their movement was accompanied by a steady clicking sound, like that of castanets," produced by tendons in their lower legs.

This herd was still streaming by when another large group of caribou approached and cut across their path. "We were treated to the sight of the two migrating through each other," wrote the scientist. "Literally thousands of animals, some files going east, some west, countermarched across the ice within yards of, but seemingly oblivious to, each other."

In their relentless advance, caribou can trample all living things. Small birds nesting on the ground suffer, but the larger ones — jaegers, various waterfowl and ptarmigan — may attempt to defend their nests. This they do by waiting until a caribou is almost on top of them, then suddenly flying up to startle the beast and turn it aside. Where thousands of caribou have passed, the land may be churned up, and in summer they can devastate areas of dry and brittle plants with their sheer numbers.

In mainland Canada, the caribou migrate over a 700,000-square-mile area. Here, as elsewhere in the North, they follow a regular pattern. By November, they are usually heading toward their winter range, the forested regions to the south that offer them a source of food in the form of lichens. Four months or so after arriving there, they begin their return journey to the tundra calving grounds, sometimes traveling as much as 500 miles to reach their destinations. So regular are the caribou in their migrations that they have worn long trails in the tundra that can be seen from the air. In the autumn of 1984, while more than 9,000 members of one herd were following their usual paths, they attempted to cross a river that was unusually swollen and were carried off by the violent current.

The midnight sun, reflected in the Arctic's Beaufort Sea, dips toward the horizon but never sets in this multiple-exposure photograph, taken at 30-minute intervals during a June night on Barter Island. The island is about 250 miles north of the Arctic Circle, at lat. 70° N.

In spring and early summer caribou may throng to certain areas, creating huge traffic jams. Kelsall, who watched the two bands crisscross, also saw northbound caribou temporarily blocked at the confluence of two rivers full of broken ice. "For several days the animals jammed the shore waiting for the water to clear," he observed. "So many and so dense were they that accurate counts were impossible. However, experienced observers made aerial estimates of 80,000 to 100,000 animals."

The females give birth in May, often with snow still on the ground. The calves can run with their mothers soon after being born, and within a week — fortified by especially rich milk — they begin foraging on their own.

Caribou feed on a variety of plants while on the move. In winter their starchy lichen diet provides them with valuable carbohydrates, essential for generating the energy they need to keep in motion, dig for lichens and maintain their body heat in the severe cold. In spring and summer they browse on the shoots of sedges, grasses, dwarf willows and birches, all high in protein — and they have even been known to eat a trampled lemming or two, or at least lick up the fat and juices. From time to time they will ferret out the carrot-shaped roots of louseworts with their fleshy lips, nudging away the soil and then yanking the plants from the ground. (More often the caribou will eat the flowering head.)

The restless character of the caribou and reindeer serves them well. By moving from area to area to feed, they reduce the likelihood of overgrazing any one spot, thus ensuring that the next time they pass, there will still be browse to eat. Overgrazing could be disastrous for them — and for humans depending on them as a principal source of food. The Alaskan Eskimos learned this the hard way. When they began hunting with guns, they killed so many caribou that they nearly depleted the herds by the turn of the century. To ensure an adequate supply of meat, 2,000 reindeer were im-

As summer wanes in Alaska, a male caribou sheds the tender skin covering its regal antlers; the shedding is a harbinger of the mating season and the migra

inter feeding grounds.

ported from Siberia. Hunting was controlled, and the wolves that preyed on the reindeer were shot. The reindeer multiplied; by 1940 there were 650,000. Predictably, overpopulation led to overgrazing and starvation. The reindeer began dying off, a few at first, soon thousands. By 1950 only 50,000 were left. And the denuded tundra, where plants take so long to grow, would take years to develop new ground cover.

Caribou have only four major natural enemies. Golden eagles and bears prey on the calves. Wolves hunt the caribou in packs and can take a high toll. But insects in summer are by far the most numerous and relentless predators. They descend in clouds on the hapless caribou in July and August; they become active when the temperature hits 50° F. Experiments have shown that in one day mosquitoes may drain as much as four ounces of blood from an animal. Driven hither and yon by such pests, a caribou may be unable to feed adequately and put on the fat needed for the winter.

To drive off the insects, caribou flick their ears, shake, twitch, shudder, leap and bound. Under severe attack, they rush to ridges, where the cool winds inhibit the pests, or take refuge in patches of snow; some partially immerse themselves in water. Often they clump together; studies have shown that many bloodsucking insects need space to buzz their target before zeroing in on it and that clustered animals make this difficult. But escape is never complete, especially from the warble and nose bot flies.

The female warble fly lays her eggs on hairs of the victim's legs. The eggs hatch in a week, and the larvae then do an incredible thing — they scissor through the skin and travel under the hide to the caribou's back. There they cut air holes and gradually grow to the size of the first joint of a man's little finger. After completing larval development, they pop out and pupate on the tundra to emerge as adults, ready to repeat the vicious cycle. So insidious are they that it is almost impossible to find a cured hide without scars or holes. Curiously, they do not infest the calves.

The nose bot fly is every bit as horrific. The female deposits her larvae in the caribou's nostrils. From the nostrils they travel to the throat and winter over in this moist, warm environment. In March their growth accelerates, and the poor animals carrying them about sniff and cough and raise and lower their heads in a desperate effort to be rid of them. Eventually the plump larvae do drop to the ground, and after passing through their pupal stage, they emerge as flies to begin a new assault on their unwilling hosts. The caribou respond to attack by pressing their noses into the grass or onto the ground in vain attempts to keep out the invaders.

The other large mammal of the tundra is the musk ox, a big, short-legged beast that lives there year round and is surprisingly agile for its size. Like the caribou and reindeer or the bison of the American prairies that it most resembles, it is a herbivore. It feeds on grasses, willows and other plants, but it is much heavier than the caribou and reindeer — a bull may weigh as much as 900 pounds. Both bulls and cows have horns, great curling ones, sharply pointed at the upturned tips, that can measure 29 inches across. Musk oxen are magnificently adapted for life on the tundra. Their bodies are covered with long, brown to black hair that forms a curly ruff on their shoulders and hangs down over their ribs almost to the ground. The undercoat consists of soft, woolly hair known as qiviut, which can be spun and woven into garments that are immensely warm yet incredibly light. This undercoat is effective insulation indeed; when the musk oxen lie down in the snow to rest or sleep, they exude so little heat that the ice crystals beneath them fail to melt.

In only one respect can musk oxen be said to have an Achilles heel — their defensive behavior. Under attack by wolves, they assemble in a line or circle; such a wall of muscle and horn served them well for eons, but in more recent times it almost proved their undoing. When hunted by men with rifles, they formed the same phalanxes and, in the process, made themselves easy targets and suffered heavy losses. Thanks to conservation efforts, they have begun to come back. Reintroduced to Alaska, they now flourish on Nunivak and Nelson Islands in the Bering Sea and at three mainland sites, where their numbers are growing phenomenally. They have been exported to Norway, Siberia and the Soviet Union's far northern Wrangell Island. Wild herds still roam northern Canada and Greenland.

For all their apparent hardiness, the tundra's animals and plants lead precarious lives. It would take little to upset the delicate balance that exists for them now. Reindeer overgrazing their pastures and musk oxen decimated by hunters with guns are only two examples of animals put in peril and later saved by human intervention. But more is at risk today; indeed, tundra itself could become radically altered if indifferently treated.

The possibility of disaster exists as interest develops in the Arctic as a source of minerals, oil and gas. There is evidence enough of what careless exploitation could lead to. Removal of ground cover produces thawing and erosion. Tracks of vehicles that rolled across tundra 40 years ago can still be seen, ugly slashes in the earth.

Even the atmosphere could threaten the ecological balance. Although the

An Arctic hare grazes amid the cotton grass in Canada's Yukon Territory. The hare's strong-legged mobility lets it roam widely in search of food, and though relatively large — up to 12 pounds — it is small enough to find shelter from winter winds on rocky tundra uplands.

air is generally very clear, there are some signs that pollution is increasing. Environmentalists were startled to discover that on a few winter days, pollution in the Far North reached levels comparable to those of suburban and rural areas of the continental United States. Ten years of research have revealed that a kind of arctic haze extends over the region. From analysis of chemicals in the air, investigators believe that much of it originates in industrial areas of the Soviet Union and Europe. North America apparently contributes little to the problem, largely because of its location and the flow patterns of the air. The lack of heavy rain and snow in the Arctic means that the pollutants are not readily washed or scrubbed from the atmosphere. When they hit the ground, they can do serious damage.

The toughest plants, those able to exist where all others cannot, would be among the first to perish — the lichens. Because they grow so slowly, they absorb substances present in atmospheric moisture and dust, and these can build up within to dangerous and even lethal levels. In Great Britain, the Netherlands, Belgium and Luxembourg, many local lichen species have become extinct, victims of factory emissions. On the tundra, lichens have accumulated radioactive substances released by nuclear test explosions over polar regions. Reindeer and caribou feeding on them picked up the radioactivity and passed it through their flesh to those creatures that fed on them in turn, including Eskimos and Lapps — to what effect, no one knows.

This hardy, delicate land — the crown Earth wears around its head — is a barometer of the physical state of the world. It must be preserved in sound condition, for if some sort of ecological disaster should occur in the lands to the south, it might be from these northern reaches that life once again starts — in the form of seeds buried deep in the frozen ground. Ω

Well-adapted to its habitat, a small white fox curls comfortably in its dense winter coat on frozen Hudson Bay. The fox preys on hares, lemmings and sea-bird colonies in the summer and caches a portion of its kill in remote rock crevices to help sustain itself through the winter.

SURVIVAL NORTH OF THE TREE LINE

Vegetation in the frigid world beyond the tree line is surprisingly varied and often striking in its beauty. No fewer than 600 species of plants survive the cruel climate of the tundra.

As the photographs on this and the following pages demonstrate, the landscape changes as it stretches northward from the coniferous forest. At first, dwarf trees and shrubs brighten the land with luminescent autumn colors. These give way to verdant meadows of grass and sedge, and finally to the polar desert, where scattered communities of moss and lichen and only a few species of flowering plants cling to life.

Variations in temperature and soil moisture are primarily responsible for the tundra's diversity. During the milder summer months, the top layer of soil thaws; plants can then grow and blossom. In the colder reaches, where temperatures remain low even in summer, the critical thawed zone may narrow to a few inches; much of it is saturated by meltwater unable to drain into the solid permafrost beneath. Colder temperatures also slow the pace at which dead plant tissue decomposes into the nutrients on which living plants depend; the scarcity of available nutrients may cause the plants to suffer from malnutrition.

These same climatic variations sometimes occur within a single region, creating markedly different growing conditions. A depression in an otherwise well-drained area of semi-arid tundra, for example, may create an ideal bog for certain moisture-loving grasses, or a hillside in a damp meadow may drain well enough to permit a grove of trees to take root. Such microclimates, like their larger counterparts, engender a splendid mosaic of life in unexpected places.

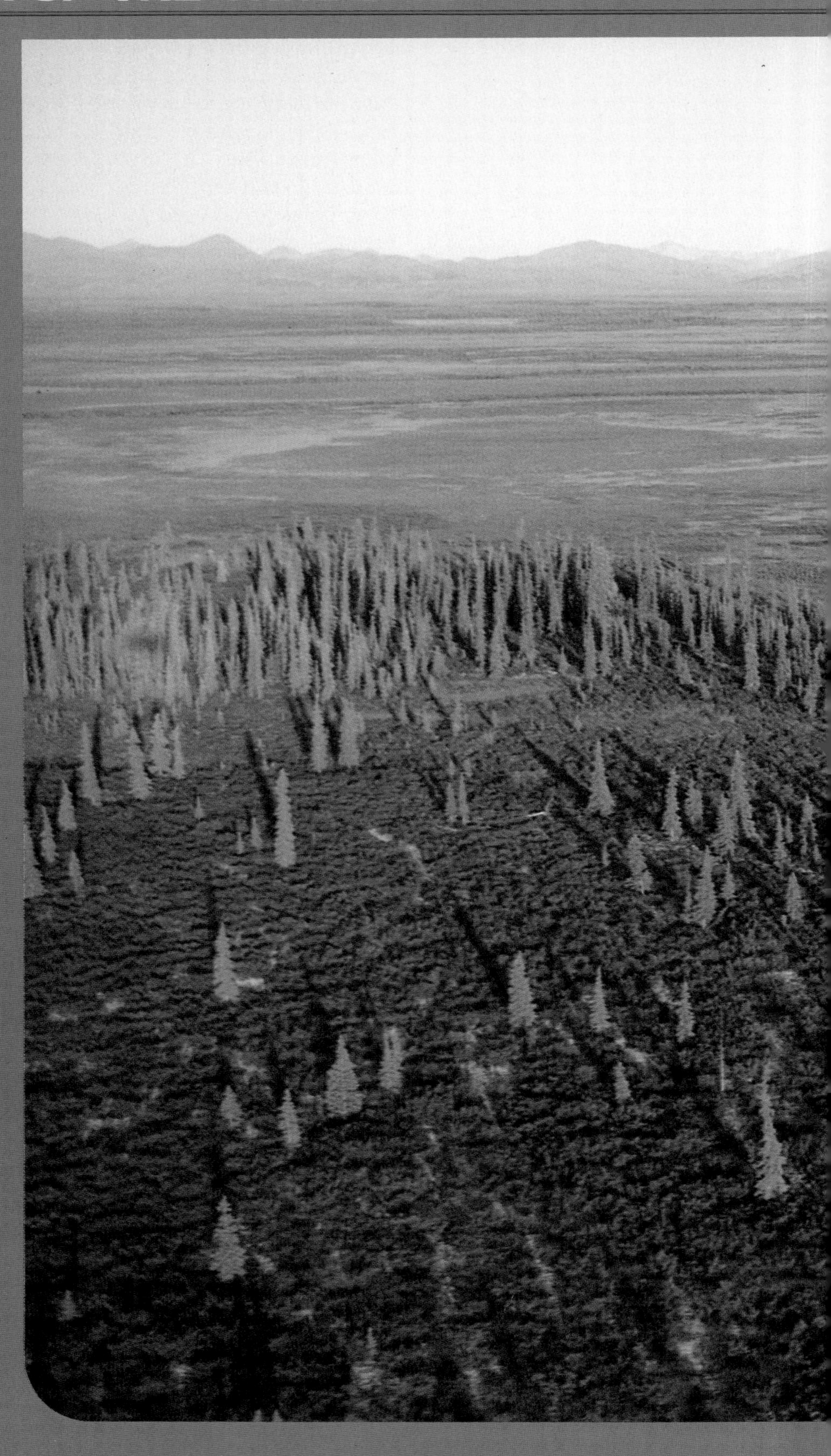

In an aerial view of Alaska's Noatak National Preserve, a grove of stunted spruces gives way to grasses and sedges that flourish in a tundra bog. The shallow pool in the distance is all that remains of a lake that once covered the bog.

Bearberry and willow shrubs brighten the rockbound shore of Canada's Hudson Bay in early autumn. Sheltered from the bay's frigid winds by boulders left behind by a glacier, the shrubs thrive in the well-drained coastal soil.

A broad meadow of cotton grass blooms near the barren foothills of the Brooks Range in Alaska. The dark slashes meandering through the meadow are low-lying areas that are too moist to support the cotton grass.

In the polar desert on Canada's Ellef Ringnes Island, thin patches of grass near seasonal ponds are the only life evident in midsummer. This part of the Arctic island supports fewer than 20 species of grass and other flowering plants.

Chapter 5

MAN, THE DUST MAKER

The dust fell from the Kansas sky on March 15, 1935, like so much coffee-colored snow. A Garden City woman sat resignedly in her living room, surrounded by her family. "All we could do," she recalled later, "was just gaze at each other through the fog that filled the room and watch that fog settle slowly, silently, covering everything — including ourselves — in a thick, brownish gray blanket." There was no relief. "When we opened the door," she said, "swirling whirlwinds of soil beat against us unmercifully. The doors and windows were all shut tightly, yet those tiny particles seemed to seep through the very walls. It got into cupboards and clothes closets; our faces were as dirty as if we had rolled in the dirt; our hair was gray and stiff; and we ground dirt between our teeth."

This storm was one of hundreds that swept across the American prairies during the 1930s, storms that blew great clouds of soil as far east as New York City. In several states the dust hung so thick that people sickened; some died of respiratory illnesses brought on by the silica in the air. Hospitals took to wrapping patients in wet sheets so they could breathe, and gauze face masks became everyday attire for citizens going about their business on the streets during a blow.

The dusters, as the storms came to be called, began in 1932, after a period of abnormally high temperatures and signs of drought. In 1933 Western weather stations reported no fewer than 179 dust storms. Thereafter the dusters steadily gained in intensity. A single storm in 1935 destroyed five million acres of wheat in Kansas, Oklahoma and Nebraska, and the soil borne aloft by the wind and transported elsewhere was estimated at twice the volume of earth excavated during the construction of the Panama Canal. Another monster storm dumped 12 million tons of dust on Chicago.

The Dust Bowl — the area most affected by wind and drought — was a vast region of 97 million acres made up of sections of Kansas, Oklahoma, Colorado, New Mexico and Texas. For three consecutive years, Amarillo, Texas, was hit by an average of nine storms a month from January to April. Wind-blown dust buried barns, houses, fields and farm equipment in drifts up to 25 feet high; after the worst storms, the sun did not shine through for several days. The toll in human misery was enormous. Hundreds of farms and ranches failed, and thousands of unemployed agricultural workers migrated westward in search of jobs. Many who stayed behind needed government relief to survive.

The Dust Bowl, whose destructive erosion continued until 1940, was the worst ecological disaster ever to hit the United States. Moreover, it

A lumbering harvester, reaping corn in an Iowa field, presents a paradoxical picture of success. Even as modern machinery and intensive farming techniques wring greater yields from the land, they tend to accelerate the exhaustion of the soil.

Automobile headlights blaze ineffectually in the eerie midday dusk of a dust storm at Amarillo, Texas, in April of 1936.

was but a single manifestation of an intensifying global crisis: The world's major croplands — most of which had originally been grasslands — were deteriorating steadily under pressure of imprudent human use. Overcultivating and overgrazing, along with cyclical drought, were leading to increasingly frequent dust storms, diminishing productivity and sometimes to widespread famine.

In fact, the harmful effects of overgrazing and overcultivating were clear and well known, though often unavoidable for practical reasons. Overgrazing removes the vital leaf area needed for photosynthesis; grass plants wither and die, leaving the soil open to wind erosion and to hard rains that pit and gully the ground. Even worse damage can be done by overcultivation. Plowing alone rips off the continuous, deep-rooted ground cover of perennial grasses, exposing all of the soil to erosion save for narrow rows of planted crops. Most of the crops are cultivated grasses such as wheat and corn; they are annuals that never have time in their one growing season to develop a dense mat of ground-stabilizing roots. The crops, force-fed with chemical fertilizers to sustain and increase productivity, rapidly exhaust the soil, making it doubly vulnerable to the forces of erosion.

These hard lessons would have to be learned again and again. But the American Dust Bowl of the 1930s did have a therapeutic effect. It sounded a loud warning that conservation efforts — ways and means of reducing the exhaustion and erosion of priceless grassland soils — would have to be adopted on a massive scale.

In the aftermath of the Dust Bowl, few American farmers cared to admit that they and their forebears had done a great deal to bring down nature's wrath. But there was ample evidence that the trouble began as soon as the prairie was plowed and planted with crops. Early settlers had left behind letters and diaries noting frequent dust storms; as early as 1855, a Kansas newspaper editor had complained that dust frequently sprinkled food like

Wind-blown dust lies piled high against a farm building in the aftermath of an April 1936 storm in Cimarron County, Oklahoma.

"condiment." Analysis of tree rings showed that severe droughts had occurred every 20 or 22 years. (Scientists speculated that the dry spells might be linked to the cycle of sunspots.)

By the time the dust storms abated, scientists of various specialties began taking a closer look at the prairies, trying to understand the root causes of the disaster. They studied soils and ground cover of every sort. Once vegetation was stripped off, fine-textured soils proved to be the most vulnerable to erosion; the richer organic soils exhibited a greater degree of an indispensable characteristic known as flocculation, which gave them staying power. Flocculation causes particles of dirt to cling together in clusters like fish eggs. Channels between the particles allow water to percolate down into the soil. But when drought sets in, the channels break down; water rides over the surface without sinking in, vegetation fails to take hold and the parched ground is left barren, prey to wind erosion.

During a blow, the wind operates like a plow, driving the heavier soil particles along the ground and lifting the lighter ones into the air. With the arrival of a polar air mass, the atmosphere becomes charged with static electricity that churns the dust into what has been well described as a "cold boil" and swirls it a mile or more high. So powerful was the build-up of static electricity during some of the worst storms that it caused automobile ignitions to fail and gave shocks to the unwary as they grabbed metal pump handles or cast-iron frying pans.

The question remained: how to reduce the disastrous soil loss? "Nature has established a balance in the Great Plains by what in human terms would be called the method of trial and error," read a 1937 report of the government-appointed Great Plains Committee. "The white man has disturbed this balance; he must restore it or devise a new one of his own." The effectiveness of grass as a ground cover led some scientists to suggest that grass be sown on some marginal croplands. The committee recommended that as many as 15 million acres of the Dust Bowl be planted again to grass.

But restoration of prairie on so grand a scale could not be achieved easily. To begin with, fences denoting private ownership stood in the way of wholesale conversion. But the basic problem was that overused land would take so long to heal. First would come weeds; then the grasses would begin to appear, but only a few species at a time. The full succession of grassland plants — according to a landmark study conducted in abandoned fields in Kansas and Oklahoma — would take years to complete. In a fascinating sidelight to their investigations, the scientists found that wild sunflowers affect the succession in the early stages; sunflower leaves and roots contain chemicals that inhibit the germination of many competing weeds. The annual grass plants that replace the weeds are dominated by the hardy *Aristida oligantha,* a species that flourishes in soils of low fertility. Finally the native perennials, led by the little bluestem, reinvade the abandoned field.

Since it was financially unfeasible to let the damaged portions of the prairies and plains return to their natural state, other conservation measures had to be devised. Barriers were needed to diminish the wind's avalanche effect on the loose, dry soil. One measure was to plant windbreaks; a crop such as sorghum, because of its height, could be used. Another measure was to create strips of grass running perpendicular to the direction of the prevailing wind. The grass would reduce wind speed along the surface and catch and hold flying dust from strips laid bare to grow crops.

But if any of these measures was to have widespread implementation, the

Replanted napier grass *(foreground)* and acacia trees *(middle ground)* encroach on barren, eroded rangeland *(background)* in an Ethiopian reclamation project. To reestablish good pasturage, grasses must be left ungrazed for at least a year — a difficult condition to impose on a country in desperate need of fodder to raise stock for food. Thereafter, herders are encouraged to harvest the forage and feed it to penned-up stock — a procedure that will curtail future overgrazing and a new round of runaway erosion.

government would have to take a hand and foster what was being called the new scientific agriculture. Chief proponent of such a service was Hugh Bennett, a South Carolinian who had seen for himself the effects of exhausted soil on the rural populations of the South. He was named director of a temporary agency set up within the Department of the Interior.

Bennett, realizing that the problems of erosion could not be solved quickly, lobbied hard for two years to bring a permanent agency into existence. Scheduled to present his views to a Congressional committee, he managed to stall until a dust storm arrived over Washington. Then, in the gathering gloom of the hearing room, he said dramatically, "This, gentlemen, is what I have been talking about!" Soon thereafter Congress approved the creation of the Soil Conservation Service (SCS). Within the same year, it passed the Soil Erosion Act, which enabled the government to launch an ambitious program for saving land.

Not least of the SCS's many accomplishments was the planting of shelterbelts and windbreaks — 220,000 trees on 30,000 farms. The trees, many of which still survive, were planted by the young men of the Civilian Conservation Corps, an adjunct of the SCS. In addition, the SCS encouraged the formation of local conservation districts, each of which would establish effective regulations for its own area.

Farmers living within these districts received instruction on how best to help themselves. Instead of cutting straight furrows with their plows, as had long been the custom in semi-arid regions, they began following the contours of the gently rising and falling land. They also built terraces that would catch and distribute the little rain their fields received. When contouring was used in combination with the building of ridgelike terraces to dam the runoff of rain water, the net effect was to increase the efficacy of a two-inch rain threefold. As a proponent of terracing described it, the water rushes back and forth between the ridges until, having no place to go, "it gets disgusted and quits," sinking into the soil.

The shortage of adequate moisture was — and remains — one of the biggest single problems in the prairies and plains as well as in grasslands around the world. Precipitation normally fluctuates from year to year and is generally most erratic in regions that receive minimal rain or snow. To compensate for irregularities, additional wells were sunk. Between 1933 and 1939, more than 700 new wells were dug in Kansas alone, doubling the number in operation. For their supply they drew on the Ogallala Aquifer, a thick underground layer of sand and gravel stretching from Texas to South Dakota and filled with runoff from the Rockies.

By the early 1940s, when the return of more or less normal precipitation brought relief from the drought and the dust, conservation measures were accepted and adopted in all of the affected areas. Conservation worked; crops flourished and crop prices rose. But in the flush of success after years of hard times, many farmers soon forgot or ignored the ecological lessons they had learned at such cost. They became impatient with the terraces, which slowed down plowing and were expensive to maintain. Some farmers plowed up the strips of grass, which they had set out to cut the force of the wind, and used the space to plant more crops. Others chopped down their windbreaks. As the pendulum began to swing away from conservation in the mid-1940s, Secretary of Agriculture Clinton Anderson felt compelled to say, "What we are doing in the Western Great Plains today is nothing

short of soil murder and financial suicide." His message was drowned by the flow of dollars that was turning poor farmers into well-off individuals. Yet it was hard to blame the farmers this time; all through the war-torn early 1940s, they were patriotically producing bumper crops for the U.S. armed forces and for America's Allies.

Predictably, drought returned to the region in the 1950s. Although the dry spell did not last as long as that of the 1930s, it proved more severe. From 10 to 16 million acres a year suffered even greater erosion than they had two decades earlier. But again the return of rain ushered in a bountiful period, and people put on the same blinders.

Then came the 1970s, another drought and still more dust storms. One of the worst storms, tracked by an earth-orbiting satellite, began on February 23, 1977, when winds roared down on the Portales area of eastern New

With a good-natured tug, a young Nuer herder asserts his authority over a willful calf. These Sudanese stock raisers leave their permanent homes for a wetland during the dry season and return with the rains.

Mexico and the southeastern corner of Colorado. By afternoon, plumes of dust had risen over both states; meeting above Texas, they formed a vast, drifting pall that covered 248,000 square miles of the country. Three days later the pall could still be seen — far out over the Atlantic.

When investigators arrived in the Portales area to measure the storm's impact, they found some plowed fields that had lost three feet of soil and others that had been inundated with dust. As much as 40 tons of soil an acre had been blown away in at least two of the local counties. One could only guess how long it would take to replace losses of such dimension. Even under ideal conditions, the forces of nature can develop only one and a half tons of soil per acre per year.

Even in the 1980s, a decade of hiatus between the worst droughts, there has been serious erosion in the Texas Panhandle, southeastern Colorado and southwestern Kansas. Large sections of marginal land best suited to grazing have been turned to crop-raising by hard-pressed farmers who have had to open up more land in order to make ends meet. The poor soil in such marginal areas is exhausted rapidly, and when it ceases to be productive, the farmers have little choice but to abandon the acreage to the forces of erosion.

Like the prairies of the United States, grasslands all around the globe have been degraded by the ignorance or neglect of farmers and stock raisers. Everywhere, dust is blowing. Australia, Asia, Africa and South America have seen seen most of their plains and savannas converted to agriculture and depleted by overgrazing and overcultivation. Often the damage has been done slowly, in barely perceptible stages. But the various steps — decreasing ground cover, repeated drought, declining water table, increasing soil erosion — eventually reduce a productive region to a desert.

According to the 1984 report of the United Nations Environment Program, as much as 35 per cent of the world's surface is threatened by desertification. The report says that to save this land for food production, the nations of the world would have to invest $4.5 billion a year for 20 years in soil conservation projects and the construction of irrigation systems.

But irrigation is no panacea. Essential though it is, irrigation has created its own havoc in several large areas of the world. Salts of various kinds occur naturally in all water. When fields are irrigated, some of the water is used by the plants, and some seeps into the soil. Evaporation leaves salt behind; so do the plants, which filter it out as they absorb almost pure moisture through their roots. Where a layer of impermeable material prevents irrigation water from soaking all the way through, the water level will rise over time; only periodic flushing can prevent capillary action from carrying increasing amounts of salt to upper layers. Once the saline water touches the roots, the plants' ability to absorb oxygen and moisture declines, and the vegetation either is stunted or dies. Upon reaching the surface, the salt dries out and lies gleaming like a killing frost. This phenomenon was chiefly responsible for what happened in the farmlands of ancient Sumeria, now Iraq; and the soil there has not yet recovered after nearly 4,000 years.

In India, inadequately planned irrigation projects have brought about the salination and waterlogging of an estimated 15 million acres of precious cropland. Another 25 million acres are at risk. The problem stems from carelessly built, unlined irrigation ditches and poor drainage. The canals leak, and the fields receive three times as much moisture as the plants

An emaciated bull wanders through a parched Niger landscape, vainly seeking food in a region stripped of greenery by a decade of drought. The famine in Ni

l much of West Africa has been exacerbated by overgrazing and the cutting of trees for firewood.

Cornfields on a Soviet collective farm roll across the steppe toward the snow-clad Caucasus Mountains. The row of trees in the middle distance is a conservation measure, planted to break the steppeland's desiccating winds.

require. And all the while, the water level inches upward. The culmination is a wet desert on which nothing will grow. "It is one of the greatest unrecognized environmental problems in the world today," says a former Indian secretary of agriculture. "We are systematically destroying our most valuable resource, and nobody is paying attention."

A more general problem, overgrazing by livestock, threatens most African grasslands. Overgrazing curtails root growth and thus lowers the plants' resistance to drought. And of course the stripping of ground cover leads to more erosion and more runoff, and usually leaves less moisture in the soil for the surviving grasses.

Most people in the damaged areas of Asia and Africa cannot afford to allow the soil to rest and recover. Since food is desperately needed, more and more land is put to use — and rapidly depleted. Perhaps as much as half of Africa is grassland, and much of it has been damaged by overgrazing, the trampling of livestock or the bite of the plow. Climate changes have removed still more land from use. The Sahara, which itself was a grassland 5,000 years ago, has been advancing southward and has engulfed 251,000 square miles of formerly productive land. In another threatened area, the Sahel region of West and Central Africa, two severe droughts within 10 years of each other have left the survivors living in misery. Here, an increase in population and improvements in farming methods induced local people to overexpand their herds and their fields, actions that put more pressure on the land than it could stand. In South Africa, the region known as the Karoo, which was once fertile grassland supporting large herds of grazing animals, is now approaching desert conditions. To the north of the Karoo, the sands of the Kalahari are advancing, while to the east, vegetation is succumbing to aridity. In an effort to save the Karoo, plowed land is now being sown with grass.

Elsewhere in Africa, some ambitious schemes to convert large areas to agriculture have led to disaster. In one of the greatest fiascos, more than three million acres of bush in Tanzania, Zambia and Kenya were torn up so that sunflowers and peanuts could be planted. The crops turned out to be pitifully small, and the effort was abandoned at a cost of $100 million.

Kenya was a perfect example of Africa's increasing agricultural problems. The country had long enjoyed a 2 per cent growth rate in agricultural production; the output that resulted provided for all of the population's food needs and produced 40 per cent of Kenya's export income. But that was changed by the expanding population, the rising cost of seed and fertilizer, and the declining productivity of overused soil. In 1976 Kenya had to begin importing food. By 1985, Kenya expected to import 1.3 million tons of food — about half its total need.

Various plans have been put forth to stop and then undo the damage to Africa's savannas. Some experts have argued that many savanna areas should be left alone to revert to their natural state. The wild animals inhabiting them could then be culled to keep their numbers stable, and the meat would be made available to the local populations.

In the Soviet Union, geography has imposed a peculiar set of problems. The best croplands — the steppes, which slant all the way across the vast country in two parallel belts — suffer from heat and aridity in the southern areas, from a short growing season in the northern portions and from the high cost of transporting crops to distant population centers. These handi-

caps, together with poor agricultural methods, have kept the productivity of the steppelands low despite the richness of the soil. Fortunately, immense areas of virgin grassland remained available well into the 20th Century. The nation continued its historic practice of barely meeting its food needs by expanding agriculture rather than improving farming techniques.

For all of these reasons, agricultural policy has long been a subject of intense debate in Moscow, and when Nikita Khrushchev rose to power in 1953, he faced a series of hard decisions. Soviet agricultural production in the year just past had scarcely surpassed a level set four decades earlier under the tsarist regime, and the populace was calling for more and better food. The cost of raising the yields of existing croplands would be astronomic, since this goal would require the construction of fertilizer factories and huge irrigation systems. The alternative — to open up still more virgin land to the east — seemed to be cheaper and more viable.

In the next six years, hundreds of thousands of pioneer farmers brought 100 million acres of new land under cultivation. In that period, Soviet grain production climbed by 50 per cent, largely because of the new lands. But at Khrushchev's fiat, the new fields were planted to grain four years in succession instead of being left fallow in the fourth year. Soon a familiar pattern appeared: Soil exhaustion accelerated, and wind erosion increased.

Two groups of agricultural experts advanced different plans for coping with the problem. One group advocated that each year a third of the Soviet wheat acreage be left fallow, either in stubble from the previous crop or planted with grass to bind the soil. Further, these experts recommended shallow plowing to reduce the levels of winter freezing and springtime runoff, and they urged that strips of wheat be protected by planting alternate strips of grass. The second group, less willing to sacrifice productivity in the cause of conservation, argued for planting corn in fallow fields between wheat crops. It also favored removing the stabilizing stubble of the harvested crop as a practical means of speeding up the planting of the next crop.

The debate came to issue in Moscow in 1961, and Khrushchev cast his decisive vote in favor of the second school of thought. His decision quickly turned out to be the wrong one.

A deep drought descended on the Soviet Union in 1963 and lasted three years. Crops failed in many state collective farms, and the new lands that Khrushchev had opened up became a disaster area. Wind erosion damaged a total of 42.5 million acres of new land, and 10 million of those acres were completely knocked out of production.

Khrushchev came under increasing political pressure. In 1963 he appointed a committee to investigate the crop failures, but he never released its findings. The next year he admitted that a portion of the new lands might better be left fallow or returned to grassland for grazing. But it was too late to help Khrushchev. For his various errors, he was ousted from the premiership by party leaders in October 1964. He was succeeded by leaders who were considerably more sensitive to the need for conservation.

Meanwhile, American scientists persevered in efforts to conserve the surviving areas of nearly pristine grasslands and to repair depleted croplands by allowing them to revert to their natural state. The grassland areas available for conservation are pathetically small; there remains less than 1 per cent of the original 400,000 square miles of tallgrass prairie. Some tallgrass is found along the unplowed right of way of railways, some in isolated pockets of prairie. The biggest portion is a 50-mile strip in the Flint Hills of eastern Kansas. Because of the limestone that lies just below the surface, the region has never been extensively plowed. Cattle have made inroads there, but they have not fatally overgrazed the native grasses.

Conservationists want to set aside 320,000 acres in the Flint Hills as the Tallgrass Prairie National Park. But ranchers have opposed the park vigorously. They argue that they love the land and that they know how to preserve it. A veteran rancher told proponents of the prairie park, "We take care of the Flint Hills better than any government agency could. We burn the prairie grasses regularly, just as the forces of nature and the Indians did before settlement, to keep our cattle in productive grazing areas. We avoid overgrazing. In general we follow sound soil-and-water conservation practices we've either learned ourselves in agricultural college or that researchers and extension folks tell us about." Many ranchers shudder to think what might happen if the Flint Hills were to become a tourist attraction. Said an octogenarian: "They'll have to run blacktop roads through it. Have to have rest stations. Have to have a place to buy pop and beer. And pretty soon the realistic part of it — the back-to-nature-just-exactly-like-God-Almighty-made-it part — *is gone.* Why can't they let it *alone?"*

Conservationists counter that they cannot afford to leave it alone, that the inroads of cattle have already jeopardized the natural balance of Flint Hills. "A range pasture," a student of the prairie has said, "is preserved prairie in the sense that it has never been plowed. But it's threadbare prairie." Such areas can be fully refurbished and restored, as several ecologists have demonstrated on a small scale. These experts believe that by bringing even the tiniest piece of prairie back to its natural state, they will gain basic information about the ecosystem as a whole — and that what they learn will directly benefit ranchers and farmers alike.

One such study area was established in the disputed Flint Hills. It is the Konza Prairie, an 8,616-acre preserve owned by the Nature Conservancy, a Virginia-based organization whose goal is to save as much ecologically valuable land as possible. The Konza is being studied by ecologists from Kansas State University with an eye to using the preserve eventually as the norm against which to measure deterioration of range and field.

Among the experiments being conducted there is one involving fire. To see what effect fire has on prairie plants and wildlife, some patches of the Konza are burned yearly, others as seldom as once a decade. Where the burning is carried out annually, the dominant grass species tend to become even more prevalent; since these species are best for cattle, ranchers burn each year to benefit their herds. Areas put to the torch every four to six years rebound with tall, luxurious grass; this burning schedule benefits wildlife by allowing time for a diversity of grasses to spread. But those sections swept by fire only once a decade begin to develop a growth of trees and shrubs that could eventually alter the area's character. The Konza scientists think that the natural burn cycle on the virgin prairie was on the order of once every three to five years.

Another restoration project is being conducted in a rather unusual place: near Chicago, in a 617-acre plot lying inside a ring formed by the huge atom smasher at the Fermi National Accelerator Laboratory. When the laboratory was commissioned in 1972, someone suggested that landscaping be

A living relic, the largest remaining patch of tallgrass prairie sprawls across the Flint Hills in eastern Kansas. The area was spared from cultivation because muc

its thin, rocky soil is unplowable. But the threat of industrial development is increasing.

A man-made tallgrass prairie flourishes inside the huge ring-shaped atom smasher at the Fermi National Accelerator Laboratory near Chicago. The planting of this 617-acre plot began in 1975 with seeds collected from small patches of prairie within a 50-mile radius of the laboratory.

Researchers harvest grass seeds from the Fermi Laboratory's model prairie. The large size of this tract allows scientists to study the interaction of the environment and its native flora and fauna.

eschewed in favor of turning the grounds into a model prairie. Planting got under way three years later with the sowing of 210 pounds of seed in a seven-acre tract. Gradually the seeded acreage has been been extended. Various plowing and planting methods have been used to determine which are most effective. Such basic prairie grasses as bluestem and Indian grass have been quick to take root and establish themselves. Grass species from abroad were introduced, and there soon appeared signs that the native plants could hold their own against the exotics. Some drought-resistant native grasses choked out the imported species, in part because their longer roots absorb subterranean moisture that the aliens' roots cannot reach. The Fermi ecologists are pleased with what they have so far accomplished, but they point out that it may take 50 to 100 years for their little prairie to approach natural species composition.

Such studies may do more than abet the restoration of prairie ecology. Researchers will save dwindling prairie plants that might yield new foods and medicines, or serve as the chemical models for synthetic products. The U.S. government puts enough stock in this possibility to maintain a Medicinal Plant Resources Laboratory in the Department of Agriculture; its job is to identify plant substances that may eventually prove useful.

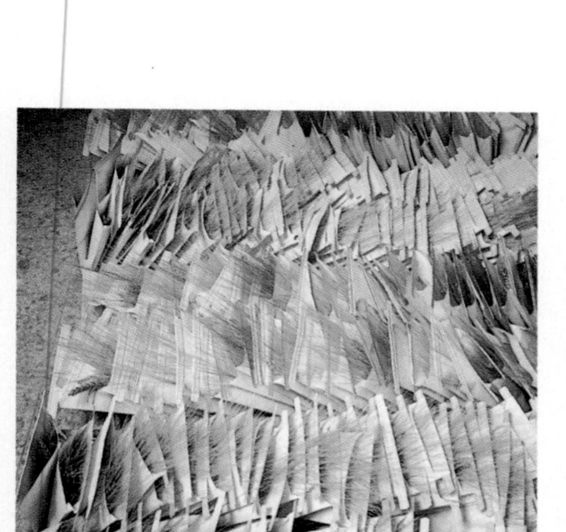

Glossy heads of barley grain bristle from sample packets filed in a gene bank in Ethiopia. Experts grade the seeds in an effort to determine and conserve the species best adapted to Ethiopia's climate and terrain.

There is a more basic reason for protecting the indigenous grass species. Together they constitute a gene pool that generations of plant breeders can draw on. Genetic tinkering has contributed largely to the enormous crops that the United States has been producing in the face of soil depletion and an increasingly polluted environment. Currently, American farmers grow about 60 million metric tons of wheat per annum, or one seventh of the world's supply — an increase of 115 per cent since 1930. And since that year, the yield of American corn has gone up an incredible 400 per cent or more, from about 10 bushels an acre to between 40 and 100 bushels an acre. About half this increase can be attributed to genetically induced improvements in the corn. And even this prodigious gain pales in the light of projections for the future. Plant geneticists believe that under ideal conditions corn production could be increased to 6,663 bushels an acre.

But the harvests have been increased at a high price. The costs of fertilizers, pesticides, herbicides and fuel for farm machinery have skyrocketed in recent years. Most important, the improved strains are short-lived. They require repeated infusions of germ plasm — fresh genetic material — to counteract new viruses and maintain the health and fertility of the seed grain. Obtaining this plasm for every kind of grain is becoming harder and harder. In the case of corn, some is derived from wild species and some from ancient cultivated varieties found in remote parts of Mexico and Central America, where corn was first domesticated more than 6,000 years ago.

The discovery in 1979 of a wild ancestor of corn in southwestern Mexico has been heralded as the "botanical find of the century." Botanists found the corn growing in three small patches that were threatened by nearby development schemes. The species possessed a unique feature: Unlike all other corn, it is a perennial. Moreover, it has the same number of chromosomes as domesticated corn, which means that it can be crossbred and thus can pass on its perennial habit. The implications are enormous. If American corn farmers did not have to plow and sow each year, they would save at least $300 million in energy costs annually — and perhaps as much as 10 times that amount. Even more significant, the plants' roots would hold the soil in

The productive power of grasslands like this Canadian wheat field offers the best hope for a solution to the mounting problem of world hunger.

place. And there is still another benefit. The wild corn is immune to four of the eight diseases that afflict ordinary corn.

The search for germ plasm is by no means limited to the last hideouts of wild species. Almost equally important to plant geneticists are the primitive varieties that farmers have been planting for centuries in outlying areas. Because the same species often differs from one district to another, the remote croplands are a rich source of genetic material. But just as wild corn is threatened by civilization, so the primitive varieties are becoming increasingly scarce as farmers abandon them in favor of the more productive strains developed by science. Worse, farmers giving up their old corn often eat the last of its seeds. And thus, as an unhappy botanist said, "the genetic heritage of a millennium can disappear in a single bowl of porridge."

Corn — which accounts for one sixth of the food eaten worldwide today — needs all the help it can be given. Genetically, the varieties of corn grown in American fields are few in number and very similar. Therefore it is theoretically possible that a virulent disease for which the corn has no built-in resistance could wipe out one season's crop and produce widespread hunger. This came close to happening in 1970, when a fungus struck corn from the Gulf of Mexico to the Great Lakes. Fifteen per cent of that year's crop was lost, and the disaster might have been compounded in years following had not fungus-resistant genes been bred into subsequent seed. At the time, 70 per cent of the corn being raised in the United States traced its ancestry to only six lines. Plant pathologist J. Artie Browning of Iowa State University, testifying before the Senate Agriculture Committee, underscored the peril of the situation, comparing America's inbred Corn Belt to "a tinder-dry prairie, waiting for a spark to ignite it."

Nor is corn the only grain so threatened. Wheat is even more prone to disease. Despite the persistent efforts to breed resistant strains, every five years or so the defenses of commercial wheat begin breaking down, and the grain must be reinvigorated with new germ plasm. Mexican farmers had to resort to six different wheats in 20 years, and American farmers used no fewer than 700 varieties in recent decades to ensure successful harvests. When stripe rust attacked American wheat in the 1960s, a wild wheat from Turkey came to the rescue. Though its strengths were unknown at the time, it has subsequently passed along to new strains its resistance to 50 disease-causing microorganisms.

In addition to developing more vigorous wheat, plant breeders want to increase the number of grains produced on a single stalk from the present 20 or 30. They have already managed to coax a stalk to bear as many as 70 grains and are now attempting to induce wheat to grow more than one head to a stalk. If they succeed, they will have elevated yields beyond their wildest dreams — by 50 to 100 per cent. But since a tall wheat plant with heavier or multiple heads will bow down of its own weight or blow over in the wind, scientists are developing shorter-stemmed varieties that will have the strength to stand up under their heavier load of kernels.

Wheat, like corn, needs continuing infusions of fresh germ plasm, and the search for wild varieties is even more difficult. In the last 40 years, 95 per cent of Greece's wild wheat has disappeared — sacrificed in favor of higher-yielding species. The situation in the Middle East, where wheat was first raised as a crop, is almost as precarious. The United Nations Food and Agriculture Organization has predicted that by the end of the 1980s, wild wheat may have vanished from the Middle East and also from Asia Minor and Ethiopia, two other prime sources of the vigorous stock that geneticists must have. In a stopgap measure, seed banks have been set up in various countries, but the germ plasm collected there cannot be kept indefinitely.

Overhanging all of these problems are some grim absolutes. Agricultural specialists declare that by the turn of the century, the worldwide limits of land available for agriculture will have been reached. The pinch is already being felt by American farmers; in the past two decades they have put into production about 58 million acres of idle land, much of it marginal in nature. At the same time, however, land that could be used for agriculture is being lost at a distressing rate — as much as 12 square miles a day, according to one estimate. Strip mining alone has consumed more than 1.2 million acres in a decade, and even after topsoil is spread over the reclaimed sections, the slow-growing grasses of the region will take years to heal the scars. But this is not all. About one million acres of cropland a year are being taken over for rural and urban development, highways, airports and reservoirs. And hundreds of thousands of acres of grassland are drying out: They are undergoing desertification.

To water their fields, many farmers have had to rely on the Ogallala Aquifer. Today it irrigates 11 million acres, but it is being depleted at an alarming rate. In a 1981 report, *Desertification of the United States,* the Council on Environmental Quality declared that while farmers may not run out of groundwater, they may soon find it too costly to use. Only a decade or so ago in Gaines County, Texas, water could be pumped for $1.50 per acre-foot. The cost has since risen to $60 because the energy used to operate the pumps has increased in cost and because the lowering of the water table by an average of 12.8 feet has made the water more expensive to obtain.

Farmers in Gaines County are removing Ogallala water twice as fast as the aquifer can replenish itself. Such a declining water table portends disaster of another sort. When more water is drawn from the ground than can be readily replaced, an aquifer's capacity shrinks as the pores that contained the liquid collapse. In western Kansas, for example, the aquifer was 58 feet thick in 1930; it is eight feet thick there now. In areas where usage is high, the ground may subside. Though no longer a part of the American grasslands, California's San Joaquin Valley offers dramatic evidence of the disruption that subsidence can cause. Here sections of the valley floor have sunk 29 feet. Deep, wide cracks and fissures have opened up, buckling highways, damaging canals, and disrupting drainage and irrigation ditches.

In the face of all this, U.S. food exports — 164 million tons in 1980 — are projected to rise to 575 million tons by the year 2000 to meet the needs of a world population that will grow from 4.8 billion to 6 billion by the end of this century. But to provide this volume of exports and to supply the domestic market as well, American agricultural productivity will have to increase by 60 to 85 per cent.

Some agricultural experts hold that the problem of world hunger can be solved by the year 2000 — that the necessary techniques and technology are already available. But the colossal cost of conservation and remedial action casts doubt on that optimistic prospect. Only two things seem certain: The conflict between urgent human needs and long-term ecological needs will surely intensify. And the pressure to exploit the world's grasslands is bound to increase. Ω

BOLD PATTERNS OF CONSERVATION

When natural grasslands are put to use as cultivated croplands, farmers inevitably strip away most of the grasses and expose the soil to water-and-wind erosion. But to slow the increasing loss of soil, the growers have adopted a number of special farming methods. Quite incidentally, these techniques create many beautiful sculptural patterns, some of which are shown here and on the following pages in aerial photographs.

Effective soil conservation, as farmers have learned, is not necessarily costly or time-consuming. Indeed, a method known as conservation tilling is generally easier and less expensive than conventional farming practices. After the harvest and when preparing the land for planting, the farmer simply leaves as much stubble as possible on the soil's surface. The stubble holds the soil together and reduces erosion.

Soil conservation is more complicated on sloping farmland, where rain is likelier to cut gullies, and in areas such as the Great Plains, where high winds carry off topsoil from the naked earth between the rows of wheat or corn. To reduce water erosion on slopes, runoff is controlled by contour tillage, a method in which a field is plowed and cultivated across a slope rather than with it, giving water less freedom to plunge downhill and gully the land. Another technique controls water erosion by breaking up long slopes with a series of terraces or ridges. The terraces hold runoff until it sinks in, rather than letting it flow uninterrupted downhill. Terracing is more expensive than contour farming, but it is also more effective. A study of Wisconsin farmland found that contoured fields lost six times as much soil as terraced fields.

Most farmlands have shallow drainage channels, which carry off surplus rain water. To prevent the channel beds from eroding, conservation-conscious farmers plant them with the best soil holder of all: grass.

Several similar techniques are used to counter other problems of erosion. Contour strip-cropping is used to reduce erosion on sloping farmlands. In buffer strip-cropping, particular crops that afford the least protection against runoff — among them corn and grain sorghum — are alternated in strips with other species of grass that hold the soil better. In wind strip-cropping, the rows of crops are also aligned perpendicular to the prevailing winds to hold down the amount of dry topsoil that is blown away.

These artful conservation methods do not stop erosion. But by curbing excessive soil loss, they improve the health and productivity of the land.

A bold sweep of golden oats and green soybeans on an Iowa farm shields the soil in two ways. These rotated crops are planted in rows perpendicular to the land's gentle slope, slowing the downhill flow of rain water. The soil is also held together by strip-cropping: Rows of oats are alternated with rows of soybeans, which are a poorer ground cover.

Amid growing crops on a Kansas farm, a newly planted field displays brown wavy lines — land roughed up by a disk plow to increase wind resistance.

A tract of naked soil in Australia is planted with grass seed in spiraling rows (with a whimsical arrow in the center) to curtail wind erosion.

Terraced and contoured against erosion, a Brazilian grassland is strip-cropped with alternating rows of soybeans and rice.

Grass-filled drainage channels, which draw off rain water and curtail erosion, link ribbons of corn and alfalfa on an Iowa farm.

ACKNOWLEDGMENTS

For their help in the preparation of this book the editors wish to thank: **In Australia:** Adelaide — Commonwealth Scientific and Industrial Research Organization (CSIRO) Division of Soils; Canberra — CSIRO Division of Wildlife and Rangelands Research. **In Italy:** Brescia — Professor Emmanuel Anati, Centro Camuno di Studi Preistorici; Milan — Marco Mairani; Rome — Florita Botts, Alessandro Bozzini, Giuditta Dolci-Favi, Fernando Riveros, Food and Agriculture Organization of the United Nations. **In the United States:** Colorado — (Fort Collins) Dr. David Coleman, Natural Resource Ecology Lab, Colorado State University; District of Columbia — Dr. Stanwyn Shetler, Smithsonian Institution; Kansas — (Hays) Dr. Joseph Raymond Thomasson, Fort Hays State University; Minnesota — (Minneapolis) Mark Heitlinger, The Nature Conservancy; Nebraska — (Lincoln) Dr. John Doran, University of Nebraska; Dr. John Holmgren, National Soil Survey Lab; North Dakota — (Woodworth) Kenneth F. Higgins, Northern Prairie Wildlife Research Center; Pennsylvania — (Philadelphia) Dr. Ara Der Marderosian, Philadelphia College of Pharmacy and Science; Washington — (Seattle) Dr. Lawrence C. Bliss, University of Washington. **In West Germany:** Hamburg — Dr. H. J. von Maydell, Susanne Schapowalow; Munich — Reinhard Künkel.

The index was prepared by Richard Mudrow.

BIBLIOGRAPHY

Books

Allen, Durward L., *The Life of Prairies and Plains.* New York: McGraw-Hill, 1967.

Barnard, C., ed., *Grasses & Grasslands.* London: Macmillan, 1964.

Bliss, L. C., O. W. Heal and J. J. Moore, eds., *Tundra Ecosystems: A Comparative Analysis.* Cambridge: Cambridge University Press, 1981.

Bloom, Arthur L., *Geomorphology: A Systematic Analysis of Late Cenozoic Landforms.* Englewood Cliffs, N.J.: Prentice-Hall, 1978.

Bosworth, Duane A., and Albert B. Foster, *Approved Practices in Soil Conservation.* Danville, Ill.: The Interstate Printers & Publishers, 1982.

Bourlière, François, and the Editors of Life, *The Land and Wildlife of Eurasia* (Life Nature Library series). New York: Time Inc., 1974.

Bradshaw, A. D., and M. J. Chadwick, *The Restoration of Land: The Ecology and Reclamation of Derelict and Degraded Land.* London: Blackwell Scientific Publications, 1980.

Brady, Nyle C., *The Nature and Properties of Soils.* New York: Macmillan, 1984.

Brandt, Herbert, *Alaska Bird Trails: Adventures of an Expedition by Dog Sled to the Delta of the Yukon River at Hooper Bay.* Cleveland: The Bird Research Foundation, 1943.

Brown, Jerry, et al., eds., *An Arctic Ecosystem: The Coastal Tundra at Barrow, Alaska.* Stroudsburg, Pa.: Dowden, Hutchinson & Ross, 1980.

Brown, Leslie:
Africa: A Natural History. New York: Random House, 1965.
The Life of the African Plains. New York: McGraw-Hill, 1972.

Butzer, Karl W., *Geomorphology from the Earth.* New York: Harper & Row, 1976.

Carleton, Ray G., *Wildlife of the Polar Region.* New York: Harry N. Abrams, 1981.

Eyre, S. R., *Vegetation and Soils: A World Picture.* Chicago: Aldine Publishing, 1968.

Eyre, S. R., ed., *World Vegetation Types.* New York: Columbia University Press, 1971.

Farb, Peter, and the Editors of *Life, The Land and Wildlife of North America* (Life Nature Library series). New York: Time Inc., 1964.

Fletcher, W. Wendell, and Charles E. Little, *The American Cropland Crisis.* Bethesda, Md.: American Land Forum, 1982.

Freuchen, Peter, and Finn Salomonsen, *The Arctic Year.* New York: Putnam, 1958.

Fuller, William A., and John C. Holmes, *The Life of the Far North.* New York: McGraw-Hill, 1972.

Gabrielson, Ira N., and Frederick C. Lincoln, *The Birds of Alaska.* Washington: The Wildlife Management Institute, 1959.

Grzimek, Bernhard, and Michael Grzimek, *Serengeti Shall Not Die.* New York: E. P. Dutton, 1960.

Hale, Mason E., *The Lichens.* Dubuque, Iowa: William C. Brown, 1969.

Hansen, Henry P., ed., *Arctic Biology.* Corvallis: Oregon State University Press, 1967.

Heady, Eleanor B., *Coat of the Earth: The Story of Grass.* New York: W. W. Norton, 1968.

Heiser, Charles B., Jr., *Seed to Civilization: The Story of Food.* San Francisco: W. H. Freeman, 1981.

Hitchcock, A. S., *Manual of the Grasses of the United States.* 2 vols. New York: Dover Publications, 1971.

Hopkins, Robert S., *Darwin's South America.* New York: The John Day Company, 1969.

Horton, Catherine, *A Closer Look at Grasslands.* New York: Gloucester Press, 1979.

Hurt, R. Douglas, *The Dust Bowl: An Agricultural and Social History.* Chicago: Nelson-Hall, 1981.

Irving, Laurence, *Arctic Life of Birds and Mammals: Including Man.* Berlin: Springer-Verlag, 1972.

Ives, Jack D., and Roger G. Barry, eds., *Arctic and Alpine Environments.* London: William Clowes and Sons, 1974.

Jenny, Hans, *The Soil Resource: Origin and Behavior.* New York: Springer-Verlag, 1980.

Jensen, William A., and Frank B. Salisbury, *Botany.* Belmont, Calif.: Wadsworth Publishing, 1984.

Jones, Evan, and the Editors of Time-Life Books, *The Plains States* (Time-Life Library of America series). New York: Time-Life Books, 1968.

Langer, R.H.M., *How Grasses Grow.* Baltimore: University Park Press, 1979.

Langer, R.H.M., and G. D. Hill, *Agricultural Plants.* Cambridge: Cambridge University Press, 1982.

McHugh, Tom, *The Time of the Buffalo.* New York: Knopf, 1972.

MacInnes, Colin, and the Editors of *Life, Australia and New Zealand* (Life World Library series). New York: Time Inc., 1964.

McNaughton, Samuel J., and Larry L. Wolf, *General Ecology.* New York: Holt, Reinhart and Winston, 1979.

Madson, John, *Where the Sky Began: Land of the Tallgrass Prairie.* Boston: Houghton Mifflin, 1982.

The Marvels of Animal Behavior (Natural Science Library). Washington: National Geographic Society, 1972.

Morris, Desmond, *The Mammals: A Guide to the Living Species.* New York: Harper & Row, 1965.

Myers, Norman:
The Long African Day. New York: Macmillan, 1972.
A Wealth of Wild Species: Storehouse for Human Welfare. Boulder, Colo.; Westview Press, 1983.

Numata, Makoto, ed., *Ecology of Grasslands and Bamboolands in the World.* London: Dr. W. Junk, 1979.

Perry, Richard, *Life in Desert and Plain.* New York: Taplinger Publishing, 1977.

Polunin, Nicholas, *Introduction to Plant Geography: And Some Related Sciences.* London: Longmans Green and Company, 1960.

Ray, G. Carleton, and M. G. McCormick-Ray, *Wildlife of the Polar Regions.* New York: Harry N. Abrams, 1981.

Richards, B. N., *Introduction to the Soil Ecosystem.* New York: Longman Group, 1974.

Risser, Paul G., et al., *The True Prairie Ecosystem.* Stroudsburg, Pa.: Hutchinson Ross, 1981.

Sears, Paul B., *Lands Beyond the Forest.* Englewood Cliffs, N.J.: Prentice-Hall, 1969.

Shelford, Victor E., *The Ecology of North America.* Urbana: University of Illinois Press, 1963.

Sigford, Ann E., *Tall Grass and Trouble: A Story of Environmental Action.* Minneapolis: Dillon Press, 1978.

Simpson, George G., *Horses.* New York: Oxford University Press, 1951.

Sinclair, A.R.E., and M. Norton-Griffiths, eds., *Serengeti: Dynamics of an Ecosystem.* Chicago: University of Chicago Press, 1979.

Smiley, Terah L., and James H. Zumberge, eds., *Polar Deserts and Modern Man.* Tucson: The University of Arizona Press, 1974.

Spedding, C.R.W., *Grassland Ecology.* London: Oxford University Press, 1971.
Sprague, Howard B., ed., *Grasslands of the United States: Their Economic and Ecologic Importance.* Ames: Iowa State University Press, 1974.
Stonehouse, Bernard, *Animals of the Arctic: The Ecology of the Far North.* New York: Holt, Rinehart & Winston, 1971.
Stores, Tracy I., *General Zoology.* New York: McGraw-Hill, 1951.
Summergill, Robert H., and Carol First, eds., *North America and the Great Ice Age.* New York: McGraw-Hill, 1976.
Swierenga, Robert P., ed., *History and Ecology: Studies of the Grassland.* Lincoln: University of Nebraska Press, 1984.
Tieszen, Larry L., *Vegetation and Production Ecology of an Alaskan Arctic Tundra.* New York: Springer-Verlag, 1978.
Tothill, J. C., and J. B. Hacker, *The Grasses of Southeast Queensland.* University of Queensland Press, 1973.
Turner, Ann Warren, *Vultures.* New York: David McKay Company, 1973.
Vankat, John L., *The Natural Vegetation of North America: An Introduction.* New York: John Wiley & Sons, 1979.
Vogel, Virgil J., *American Indian Medicine.* Norman: University of Oklahoma Press, 1970.
Watts, May Theilgaard, *Reading the Landscape: An Adventure in Ecology.* London: Macmillan, 1970.
Weaver, J. E., *North American Prairie.* Lincoln, Neb.: Johnsen Publishing, 1954.
Weiner, Michael A., *Earth Medicine-Earth Food: Plant Remedies, Drugs and Natural Foods of the North American Indians.* New York: Macmillan, 1972.
Whittaker, Robert H., *Communities and Ecosystems.* London: Macmillan, 1970.
Wolf, Eric R., *Sons of the Shaking Earth.* Chicago: University of Chicago Press, 1959.
Wright, Henry A., and Arthur W. Bailey, *Fire Ecology: United States and Southern Canada.* New York: John Wiley & Sons, 1982.
Zwinger, Ann H., *Land Above the Trees: A Guide to American Alpine Tundra.* New York: Harper & Row, 1972.

Periodicals

Batie, Sandra S., and Robert G. Healy, "The Future of American Agriculture." *Scientific American,* February 1983.
Bell, Richard H. V., "A Grazing Ecosystem in Serengeti." *Scientific American,* July 1971.
Booth, W. E., "Revegetation of Abandoned Fields in Kansas and Oklahoma." *American Journal of Botany,* May 1941.
Campbell, Lise, "An Abundance of Acacias." *Swara* (Nairobi, Kenya), January-February 1981.
Cargill, S. M., and R. L. Jeffries, "The Effects of Grazing by Lesser Snow Geese on the Vegetation of a Sub-Arctic Salt Marsh." *Journal of Applied Ecology,* 1984.
Clayton, W. D., "Evolution and Distribution of Grasses." *Annals of the Missouri Botanical Garden* (St. Louis), 1981.
Coupland, Robert T., "The Effects of Fluctuations in Weather upon the Grasslands of the Great Plains." *The Botanical Review* (New York, N.Y.), May 1958.
Cross, Michael, "UN Admits Failure to Halt Deserts." *New Scientist,* May 10, 1984.
Everett, K. R., "Wetlands of the World: Summer Wetlands in the Frozen North." *Geographical,* October 1983.
"Famine: An African Nightmare." *Newsweek,* November 26, 1984.
Farney, Dennis, "The Tallgrass Prairie: Can It Be Saved?" *National Geographic,* January 1980.
Gibbons, Boyd, "Do We Treat Our Soil Like Dirt?" *National Geographic,* September 1984.
Guthrie, Russell D., "Recreating a Vanished World." *National Geographic,* March 1972.
Harlan, Jack R., "The Plants and Animals That Nourish Man." *Scientific American,* September 1976.
Hayes, Harold, "Rhythms of the Serengeti: A Cautionary Tale." *GEO,* April 1981.
Johnsgard, Paul A., "Courtship on the Plains." *GEO,* May 1982.
Kellogg, Charles E., "Soil." *Scientific American,* July 1950.
Kelsall, John P., "The Migration of the Barren-Ground Caribou." *Natural History,* August-September 1970.
King, John A., "The Social Behavior of Prairie Dogs." *Scientific American,* October 1959.
"Kingdom of the River Rats." *Life,* December 1981.
Lockeretz, William, "The Lessons of the Dust Bowl." *American Scientist,* September-October 1978.
Lott, Dale F., "Bison Would Rather Breed than Fight." *Natural History,* August-September 1972.
McNaughton, Samuel J., "Serengeti Migratory Wildebeest: Facilitation of Energy Flow by Grazing." *Science,* January 9, 1976.
"Managing the Inland Sea." *Science,* May 18, 1984.
Myers, Norman, "A Farewell to Africa." *International Wildlife,* November-December 1981.
Porsild, A. E., C. R. Harington and G. A. Mulligan, "Lupinus Arcticus Wats. Grown from Seeds of Pleistocene Age." *Science,* October 6, 1967.
Poulton, Robin, "Cooperation against Drought." *Geographical,* October 1983.
Raven, Peter H., and Daniel I. Axelrod, "Angiosperm Biogeography and Past Continental Movements." *Annals of the Missouri Botanical Garden* (St. Louis), 1981.
Sauer, Carl O., "A Geographic Sketch of Early Man in America." *Geographical Review,* 1944.
Thomsen, Dietrick E., "The Lone Prairie." *Science News,* October 16, 1982.
Trudell, Jeanette, and Robert G. White, "The Effect of Forage Structure and Availability on Food Intake, Biting Rate, Bite Size and Daily Eating Time of Reindeer." *Journal of Applied Ecology,* 1981.
Wilson, Roger E., and Elroy L. Rice, "Allelopathy as Expressed by *Helianthus Annuus* and its Role in Old Field Succession." *Bulletin of the Torrey Botanical Club,* (Lancaster, Pa.), September-October 1968.
Zahl, Paul A., "Portrait of a Fierce and Fragile Land." *National Geographic,* March 1972.

Other Publications

Britton, Max E., ed., "Alaskan Arctic Tundra." Arctic Institute of North America, Technical Paper No. 25, September 1973.
Brown, Jerry, ed., "Ecological Investigations of the Tundra Biome in the Prudhoe Bay Region, Alaska." Biological Papers of the University of Alaska, Special Report No. 2, October 1975.
Bunnell, F. L., S. F. MacLean Jr. and J. Brown, "Barrow, Alaska, USA." International Biological Program, October 1975.
Cole, Monica M., "Biogeography in the Service of Man." An Inaugural Lecture, Bedford College, University of London, October 28, 1965.
"Desertification: An Overview." United Nations Conference on Desertification, Nairobi, Kenya, Aug. 29-Sept. 9, 1977.
Eckholm, Erik, and Lester R. Brown, "Spreading Deserts: The Hand of Man." World Watch Institute Paper No. 13, August 1977.
The Global 2000 Report to the President: Entering the Twenty-First Century. U.S. Council on Enviromental Quality, The Technical Report, Vol. 2, 1980.
Olson, James E., "The Effects of Air Pollution and Acid Rain on Fish, Wildlife, and Their Habitats: Arctic Tundra and Alpine Meadows." U.S. Environmental Protection Agency, Air Pollution and Acid Rain, Report No. 8, June 1982.
"Proceedings of the Circumpolar Conference on Northern Ecology." Ottawa: National Research Council of Canada, 1975.
Reimers, E., E. Gaare and S. Skjenneberg, eds., "Activity Pattern: The Major Determinant for Growth and Fattening in Rangifer?" Proceedings of the Second International Reindeer-Caribou Symposium, Røros, Norway, 1979.
Sheridan, David, "Desertification of the United States." U.S. Council on Environmental Quality, U.S. Government Printing Office, 1981.
Shetler, Stanwyn G., "Plants in the Arctic-Alpine Environment." Washington: Smithsonian Institution, Publication No. 4584, 1964.
"Soil Conservation Districts for Erosion Control." U.S. Department of Agriculture Soil Conservation Service, Miscellaneous Publication No. 293, no date.
Tikhomirov, B. A., "The Interrelationships of the Animal Life and Vegetational Cover of the Tundra." U.S. Department of Commerce, Springfield, Va., 1966.
White, Robert G., "Foraging Patterns and Their Multiplier Effects on Productivity of Northern Ungulates." OIKOS, Copenhagen, 1983.
Whyte, R. O., T.R.G. Moir, and J. P. Cooper, "Grasses in Agriculture." Food and Agriculture Organization of the United Nations, Agricultural Study No. 42, 1959.

PICTURE CREDITS

Credits from left to right are separated by semicolons, from top to bottom by dashes.
Cover: Tomas Sennett. 6, 7: Roland and Sabrina Michaud, Paris. 8, 9: Loren McIntyre. 10, 11: Ron Ryan/Photographic Agency of Australia, Elwood. 12, 13: © Reinhard Künkel, Munich. 14, 15: © Robert P. Carr/Natural Science Photography. 16, 17: Pat Morrow/First Light, Toronto. 18: Egyptian Expedition of the Metropolitan Museum of Art, Rogers Fund, 1930. 20, 21: © Jim Brandenburg. 22, 23: Map by Lloyd K. Townsend Jr., © Homolosine projection, courtesy of Department of Geography at the University of Chicago. 24, 25: Jean C. Prior. 26, 27: Chart by Frederic F. Bigio from B-C Graphics except fossil photographs by Dr. Joseph R. Thomasson and maps by Dr. Christopher Scotese. 28, 29: Steenmans/ZEFA, Düsseldorf. 30: © 1984 Jim Brandenburg/Woodfin Camp & Associates. 31: Illustration by Lloyd K. Townsend Jr. 32, 33: Jack R. Harlan, illustration from "The Plants and Animals that Nourish Man," *Scientific American,* September © 1976, p. 96. 34: © James P. Jackson/Photo Researchers, Inc. 36: NASA. 38, 39: © Reinhard Künkel, Munich; © François Gohier, Jurançon, France. 40: Steven C. Wilson/ENTHEOS. 41: Christina Loke/Photo Researchers, Inc. 42, 43: © 1980 Jim Brandenburg/Woodfin Camp & Associates; © Rick Smolan/Woodfin Camp & Associates. 44: Tony Brandenburg — M. P. Kahl/Bruce Coleman Inc. 45: Steven C. Wilson/ENTHEOS. 46, 47: © 1982 Chuck Fishman/Woodfin Camp & Associates; Jon Farrar. 48: © Jim Brandenburg. 50, 51: © René Burri/Magnum, Paris. 52, 53: Carl Kurtz; © Jim Brandenburg (4); Stephen J. Krasemann/DRK Photo; © 1984 Jim Brandenburg/Woodfin Camp & Associates — drawings by Frederic F. Bigio from B-C Graphics. 56: Carl Kurtz; Ruffier-Lanche/Jacana, Paris. 57: © 1983 Robert Carlyle Day/Photo Researchers, Inc.; © 1983 Jeff Lapore/Photo Researchers, Inc.; © Jim Brandenburg. 59: © 1973 Lowell J. Georgia/Photo Researchers, Inc. 61: © Jim Brandenburg. 62: © 1980 Tom McHugh/Photo Researchers, Inc. 64, 65: © Jim Brandenburg; Steven C. Wilson/ENTHEOS. 66, 67: © Jim Brandenburg. 68: Carl Kurtz; Perry Shankle Jr. 69-72: Steven C. Wilson/ENTHEOS. 73: © 1983 Leo Touchet/Woodfin Camp & Associates. 74, 75: Steven C. Wilson/ENTHEOS. 76: Dr. David C. Coleman and R. N. Ames/National Resource Ecology Lab, CSU, Fort Collins; Steven C. Wilson/ENTHEOS. 77. © Jim Brandenburg. 78-81: Steven C. Wilson/ENTHEOS. 82: Dr. E. R. Degginger, FPSA. 85: Ron Ryan/Photographic Agency of Australia, Elwood. 86, 87: © Reinhard Künkel, Munich. 88, 89: © 1981 Stephen Green-Armytage. 91: W. Garst/Tom Stack & Associates. 92, 93: Ron Ryan/Photographic Agency of Australia, Elwood — © L. H. Newman/NHPA, Ardingly, Sussex. 94: © Kjell B. Sandved. 95: Ron Ryan/Photographic Agency of Australia, Elwood. 96, 97: © 1981 Medford Taylor/Woodfin Camp & Associates. 98, 99: Arthus-Bertrand/© Peter Arnold, Inc. 100: © Sven-Olof Lindblad/Photo Researchers, Inc. 101: © Jim Brandenburg. 102-105: © Reinhard Künkel, Munich. 106: Lowell Georgia. 108: Steven McCutcheon/Marka, Milan. 109: Fred Bruemmer, Montreal. 110, 111: L. C. Bliss. 112: © Al D. Smiley/Peter Arnold, Inc. 113-115: Botanical drawings by Andie Thrams. 116: © 1976 Tom Branch/Photo Researchers, Inc. — Fred Bruemmer, Montreal. 117: Mark McCann — Kevin Schafer/Tom Stack & Associates. 118: Fred Bruemmer, Montreal. 119: © John deVisser, Ontario; © 1983 Steven C. Kaufman/Peter Arnold, Inc. 120: Fred Bruemmer, Montreal. 121: Varin-Visage/Jacana, Paris. 122: Brian Hawkes/Jacana, Paris. 123: © Brian Milne/First Light, Toronto. 124, 125: © 1983 R. Dennis Wiancko/Travelling Image Company. 126, 127: Helen Rhode. 128: © Brian Milne/First Light, Toronto. 129: Fred Bruemmer, Montreal. 130, 131: Tim Thompson. 132, 133: Bob and Clara Calhoun/Bruce Coleman Inc. 134, 135: © 1983 R. Dennis Wiancko/Travelling Image Company. 136, 137: L. C. Bliss. 138: © Craig Aurness/West Light. 140, 141: Library of Congress. 142: Florita Botts/F.A.O., Rome. 144, 145: Robert Caputo. 146, 147: © Philippart de Foy/Explorer, Paris. 148, 149: © Burt Glinn/Magnum Photos, Inc. 152, 153: © Jim Brandenburg. 154: © Dan McCoy/Rainbow — © Jim Brandenburg. 155: Florita Botts/F.A.O., Rome. 156, 157: © 1978 John deVisser/Black Star. 160, 161: Tom C. Cooper. 162-165: © Georg Gerster/Photo Researchers, Inc. 166, 167: © Nicholas DeVore III/Bruce Coleman Inc. 168, 169: © Georg Gerster/Photo Researchers, Inc.

INDEX

Page numbers in italics refer to illustrations or illustrated text.

Pepcid®
(Famotidine)

Tablets, 20 mg and 40 mg
Oral Suspension
Injection and Injection Premixed

995011(1)-06-PEP